"Alison and Leanne have given us a gift in their work on this book. Many of us can see ways in which change needs to happen in the world around us. This book shares an insightful process that has widespread application which would support us becoming involved and more confident in our activism. The clarity of their writing impressed and moved me, and means this book is likely to have an audience in many areas of activism."

Piers Worth, *Visiting Professor, Centre for Positive Psychology, School of Human and Social Sciences, Buckinghamshire New University*

A Toolkit for Effective Everyday Activism

This book examines how everyday activists can enhance their effectiveness.

Alison Rogers and Leanne Kelly unpack theories from the social sciences to help find meaning, explain these feelings of inertia, and provide strategies to overcome them. Through lessons learned over their careers as evaluators in non-profit organisations, Kelly and Rogers provide tools and strategies for measuring, improving, and sharing the effectiveness of planet-saving activities. They draw upon interviews with everyday people who are contributing to change in their homes, community groups, workplaces, and social settings to understand how they motivate and encourage others. The book concludes with a realistic look at individual expectations and focuses on how to prioritise self-care to ensure that activists can keep contributing in a way that maintains their wellbeing and balance.

A Toolkit for Effective Everyday Activism empowers people to use theory, research, and practical tools to leverage their power so they can make the maximum contribution possible and sustain their efforts over the long term. It will be a great resource for individuals working and volunteering in community groups, NGOs, and non-profit and corporate organisations.

Alison Rogers has worked in public health for over 20 years. She spent over 15 years of this time working in non-profit organisations including as an internal evaluator and an external evaluator. She has a PhD from the University of Melbourne's Centre for Program Evaluation that investigated evaluation advocates in non-profit organisations. Alison also has a Master of Evaluation and a Master of Public Health among other tertiary degrees in philosophy, health, education, and science. She currently holds an Honorary Research Fellow position with the Assessment and Evaluation Research Centre at the University of

Melbourne. She has co-authored a book on internal evaluation and published 15 scholarly papers. Alison is the secretary on the executive committee of her local environmental networking group and writes book reviews and articles for their e-publication. She has recently opened a bookshop to foster a community around reading and writing.

Leanne Kelly has spent nearly 20 years working in community development and social service non-profit organisations across five continents. She has worked for non-profit organisations in a broad range of disciplines (from housing and emergency services to child protection and peacebuilding) with the majority of her roles focused on evaluation. Leanne has also worked as an external evaluator for non-profit organisations, most recently in Myanmar. She has a PhD from Deakin University on evaluation in small international and community development non-profits and has published 40 scholarly papers and two books: *Evaluation in Small Development Non-Profits* and *Internal Evaluation in Non-Profit Organisations* (co-authored with Dr Alison Rogers) She is currently a Postdoctoral Research Fellow at Deakin University and is the National Evaluation Advisor at the Australian Red Cross.

A Toolkit for Effective Everyday Activism

Alison Rogers and Leanne Kelly

Routledge
Taylor & Francis Group
LONDON AND NEW YORK

earthscan
from Routledge

Designed cover image: Leah Samour at Land of Lulu | Original art of earth showing interconnections between individuals

First published 2025
by Routledge
4 Park Square, Milton Park, Abingdon, Oxon OX14 4RN

and by Routledge
605 Third Avenue, New York, NY 10158

Routledge is an imprint of the Taylor & Francis Group, an informa business

The electronic version of this book was funded to publish Open Access through Taylor & Francis' Pledge to Open, a collaborative funding open access books initiative. The full list of pledging institutions can be found on the Taylor & Francis Pledge to Open webpage.

British Library Cataloguing-in-Publication Data
A catalogue record for this book is available from the British Library

ISBN: 978-1-032-36824-5 (hbk)
ISBN: 978-1-032-36825-2 (pbk)
ISBN: 978-1-003-33398-2 (ebk)

DOI: 10.4324/9781003333982

Typeset in Times New Roman
by Taylor & Francis Books

Contents

Acknowledgements

We would like to thank the following people for their review of chapters: Dianne Beatty, Sharlene Hindmarsh, Dr Catherine Malla, Dr Piers Worth, Dr Natalie Fraser, the contributors who provided feedback on early drafts, and several anonymous reviewers. Extra special thanks to Tamsin Ramone and James Brittain for reviewing the book in its entirety. Thank you to artist Leah Samour from Land of Lulu[1] in Portland, Oregon for providing the original cover artwork. Last but not least, we thank the 46 everyday activists who generously shared their insights and experiences with us; we hope these pages do you and your work justice.

Note

1 https://www.landoflulu.com/

List of interviewees

This book is based on the experiences and insights of everyday activists. Throughout 2022, we interviewed 46 people from a range of countries including Australia, Papua New Guinea, Bangladesh, France, United Kingdom, Fiji, Kenya, Nepal, South Sudan, Canada, Palestine, Myanmar (Bamar and Rohingya), Brazil, China, Indonesia, India, Aotearoa, New Zealand, Pakistan, and the Philippines. These everyday activists are also working and stay-at-home parents; Australian Aboriginal and Torres Strait Islander people (5); retirees; refugees and former refugees (6); people with disabilities (6); LGBTQIA+ people (4), researchers; students; freelancers; small eco-business owners; and non-profit, for-profit, and government staff members and volunteers. The youngest was still at high school and the oldest was 92 years old. Of the activists, 31 identified as female and 15 as male. While we did not specifically ask about religious affiliations, over a third of interviewees raised their alignment to religions including Hinduism, Islam, Catholicism, Christianity, and Buddhism. Some of the interviewees come from our own social networks but predominantly they were referred to us by others and invited by email to participate. We received ethics approval from Deakin University to conduct the research using these methods.

We found the experiences of these everyday activists to be insightful and often inspirational. We learnt something new from each contributor and appreciated their willingness to give their time and share their experiences. We have included their name, or pseudonym in some cases, at their request, along with a short description of their actions below. As we draw upon their quotations and expertise throughout the book, we encourage the reader to return to this outline of contributors at any time to understand more about the context in which the respondent is working.

Adrian

Adrian's passion for life is evident from his loud laugh and straight-talking colloquial approach. He is an elderly white Australian with a thick white beard and long hair tied into a ponytail. He lives in government subsidised accommodation for older people in Queensland. Adrian speaks articulately and confidently, drawing upon his professional background in clinical and organisational psychology. He imparts wisdom from a long history of activism over many different settings and time periods. He currently volunteers with a local group he co-founded to advocate at a weekly market stall. They lobby for policies on climate change, promote permaculture, protest coal mines, and support community members to engage with political candidates. Adrian openly and with humour discusses the wide range of emotional responses that result from his efforts, accepting that there are both positives and negatives, but always focuses on what he needs to do to protect the future for his grandchildren and their grandchildren. Adrian is particularly proud of his picketing, letter writing, cold calling, and leaflet sharing efforts as these have contributed to banks, insurance companies, and other service providers withdrawing their investments in coal companies.

Alex

Alex is a white Australian in her mid-30s with curly auburn hair. She lives with her partner and rescue dogs in the outer suburbs of Melbourne. Initially quiet, reserved, and downplaying her actions, over the course of the interview Alex shares examples of how her determination and persistence has resulted in changes in her community and workplace. Inspired after watching documentaries about waste, Alex invests a great deal of time and energy doing everything she can to minimise her personal impact on the planet. She buys things with a lower waste profile, refills multi-use bottles, separates her waste, and contributes to a local community garden. Alex helped initiate a seed library to encourage others to tap into the social support and community spirit that gardening provides. In her professional career in the health sector, Alex connects with other professionals at conferences to see how sustainability can be applied in the workplace. In her previous workplace, starting with collecting food scraps for a worm farm, she developed a recycling station system that was subsequently adopted by over 100 branches of the organisation.

Alicia

Alicia is a middle-aged self-assured and confident woman who works in the corporate banking sector in Papua New Guinea as the head of a department focused on support services. She grew up in a very poor household in Fiji with paternal grandmother and aunts and is of mixed parentage (her father is Indian Muslim and mother indigenous Fijian). Through a scholarship, Alicia was able to obtain qualifications overseas that opened employment opportunities in the construction industry and then subsequently the banking sector in Fiji and Papua New Guinea (PNG). Concerned about the extent of domestic violence and homelessness in PNG, Alicia approached her management with a business case to convert vacant properties into safe houses. After multiple setbacks, she eventually obtained permission and began networking through women's coalitions to source the expertise required for design and implementation. The project evolved into a private, public, and civil society partnership to support survivors of family and sexual violence called Bel isi PNG.[1] Alicia also advocates for legislative change through a separate foundation in relation to family violence and sorcery accusation related violence, particularly for women from rural areas. She took great satisfaction in seeing her organisation adopt a policy on family and sexual violence and witnessing the educational and employment outcomes of individuals she has assisted through the project.

Alyssa

Alyssa is a first-generation English and Canadian Australian. While navigating debilitating health challenges, Alyssa channels her energy not just into caring for her growing family (including newborn twins), but also into a small eco-business[2] and non-profit organisation. A committed wildlife activist, Alyssa has spent many years contributing to campaigns against shooting of native waterbirds, quail, and parrots. Beyond avian life, she shows her unwavering commitment to wildlife conservation by spearheading a kangaroo advocacy not-for-profit, the Victorian Kangaroo Alliance,[3] committed to drawing attention to the often overlooked plight of the world's most exploited land-based wildlife. Always keen on fostering grassroots change, Alyssa also offers community education on how people can minimise their environmental impact after death and become post-mortem eco-warriors.

Aruna

Aruna is a middle-aged Nepalese woman and works for an international non-government organisation. Her passion and enthusiasm are infectious and her ability to channel that energy with focus is admirable. Aruna has an undergraduate degree in science and holds a double Master's degree in International Relations and Business Administration. She mentors young people from around the world and trains other people to be youth mentors. She works with schools to develop curricula and deliver sessions to help young people grow their self-confidence and self-esteem. Aruna helps them connect with each other to demonstrate how we can build a peaceful world. In the interview, she recalls how satisfying it was to see angry, traumatised child soldiers in Nepal transform into humble young people interested in finding employment and gaining an education. Aruna provides an excellent example of how to pour energy into a specific area over the long term whilst maintaining a mindset of continuous learning and improvement.

Ben

Ben is a reflective person who questions everything. He applies concerted effort to understand whether his advocacy efforts are making a difference and how he can be more effective. Ben is an early middle-aged white Australian man with striking dark features and black spectacles living in inner-city Melbourne. He writes music and plays in a band as a hobby. Throughout his career in international development, he has worked for organisations focused on public health, human rights, humanitarian and refugee law, emergency management, and conflict. He currently holds an advocacy position in an international non-government organisation. Beyond his professional role, Ben takes an active interest in progressive politics, climate change, the environment, animal rights, and gender and LGBTQIA+ rights. Ben has been vegan for many years and facilitates online and face-to-face groups that share knowledge about the topics he is passionate about. Reflecting upon his career so far, Ben appreciates how satisfying a win in the advocacy space can be; he was a part of Australia's delegation to the United Nations that saw the ratification of an arms treaty in 2014.

Caroline

Caroline is a French woman who lives in Australia with her Spanish husband. She is middle-aged with blonde flowing curls and works as a financial manager in a corporate setting. Using her own initiative, Caroline created a small group called the *green team* in her workplace. This project resulted in recycling bins being incorporated into the waste management plan, a battery recycling project, and a workplace garden featuring native plants. The garden was self-funded and plants were purchased using fundraising profits from workplace catered breakfasts and recycled bottles and cans. Caroline applied her finance skills to analysing their efforts and demonstrated how much money they saved each month. She converted the volumes of materials into the equivalent number of bathtubs and sent this out via email so her colleagues could visualise the impact of their behaviour change. Caroline leaves the city and connects with nature as a way of recharging her reserves of energy. Reminding herself why she cares so much about protecting the natural environment for her nephews provides a great example of self-care.

Clariana

Originally from Brazil, Clariana is a young woman who lives and works in Canberra, Australia. While managing a chronic autoimmune disease, at the time she was interviewed for this book she was also studying for a Master's in International Development and working in the justice program of an international non-government organisation. Her work involved supporting employers to identify what they can do to recruit and retain people with experience of the justice system. Additionally, Clariana is passionate about promoting more realistic conceptions about the body image of women. She spends time discussing the issue among friends and family and sharing on social media. She has changed the perspectives of many people with whom she interacts.

Daisy

Daisy is a Filipina who established a not-for-profit organisation in Papua New Guinea called FemiliPNG[4] to fight against family violence in vulnerable communities. She forms alliances and networks across PNG with others who are fighting against family and sexual violence and focuses on ensuring that help and services reach the vulnerable in her community. She was previously a social worker for many large international non-government organisations but the idea for FemiliPNG came because she wanted to focus on both the medical and psychological care for survivors of gender-based violence. The organisation enables people to find accommodation, legal assistance, and other practical support. The organisation has recently been recognised by the PNG government as being an effective community service organisation. This acknowledgement comes with a large increase in ongoing funding and higher reputational status among other international non-government organisations.

Darrell

Darrell is a white Australian in his early 90s. He was born during the Great Depression and grew up questioning the underlying reasons for poverty and war. Initially working in the steel industry, in 1949 he joined the Australian merchant navy and became a union activist and organiser. During the Cold War and whilst serving in the merchant navy, he collected over 600 signatures for the Stockholm and Helsinki appeals against the proliferation of nuclear weapons. He then joined a political movement to continue his anti-war movement activities. When he retired and relocated to a community on the east coast of Australia, he renewed his activity in the environmental movement. Observing that there were numerous small action groups, Darrell was the founding member of an overarching organisation called EcoNetwork Port Stephens.[5] The non-profit members-based networking organisation has now been in existence for 30 years and continues to advocate on behalf of a diverse group of affiliates on local grassroots issues that threaten environmental sustainability. Although he no longer participates in formal meetings, Darrell provides an advisory role from his room in the local aged care facility. He takes any opportunity to contribute and network members actively seek out his wise counsel.

Delphine

Delphine is an Aboriginal and Torres Strait Islander woman living in Far North Queensland. She has an infectious laugh and can find humour in even the most challenging of circumstances. She is an Elder in her community, has a Master's in Public Health, a Master's in Music Therapy, and works in a remote Aboriginal community. She was inspired to be an activist by her family, with strong family and community activist influences. Eddie Mabo, the famous Australian activist for Aboriginal and Torres Strait Islander People's land rights, was a member of her community. She frequently attended meetings where Mabo and her father discussed human rights and the role of institutions such as the United Nations. Her activism career began when she became involved in establishing culturally appropriate drug and alcohol rehabilitation. Delphine also worked as a director for a local community radio station and helped design and implement a strategy for building a peaceful community with her colleagues and friends. This was a significant achievement as it led to one of the first cultural awareness programs for white police and the employment of Aboriginal community police officers in her community, which became a model for other programs across Australia.

Dia

Dia's passion for her work in her hometown of Mumbai is compelling. She speaks with confidence, dignity, and authenticity that hooks others to join her network of collaborators. Dia is an excellent example of someone doing activism well. She is continually honing her approach, empowering her crew, and widely delegating responsibility. She knows the work is never ending, but she takes every opportunity to report and celebrate wins with her community of colleagues, supporters, and recipients. In addition to her work as a nutritionist and holistic health coach, Dia is a member of the all-women-run not-for-profit Angels Link Foundation.[6] Angels Link has been adding value through running events and projects for other charitable organisations in Mumbai for over 15 years. The organisation has an innovative method of operation whereby the 15 core members take turns to run projects and each take on a certain amount of the fundraising duties. This provides members with clear responsibilities and a sense of achievement without overburdening. While Dia is strongly invested in her charitable work, she knows she cannot do it alone and cannot do it forever. This foresight and her focus on self-care, for herself and others, helps her actively engage collaborators and seek to inspire young people to sustain momentum into the future.

Domuto

Having first-hand experience of a war-torn childhood, Domuto wants better for his compatriots in South Sudan. Rather than twisting his trauma into bitter victimisation and feeding into anti-social behaviours in attempts to avenge injustices meted out against him, Domuto is fighting for peace. For the past decade he has been advocating for peace and youth rights across the country. He works at the grassroots level for a small non-profit organisation where he uses sport as a peacebuilding conduit to bring male and female young people from different ethnic groups together. As a local man with a visible presence in the community, Domuto has become known and respected as a mediator for peace.

Ekawati

Ekawati is a force to be reckoned with. She is a fiercely determined development practitioner and educator with a PhD focused on livelihoods and economic access for people with disabilities in Indonesia. Growing up in a small town as a deaf Chinese Indonesian, education and work opportunities for persons with disabilities were limited. Eka travelled to Jakarta after high school to learn American sign language and attended Gallaudet University for a year. She then transferred to University of Maryland where she was exposed and connected to students' grassroot activism and volunteers of public interest groups. Since those formative years, Ekawati has embarked on a global journey, working to advance inclusive education and disability policies, with a special focus on her homeland of Indonesia.

Elise

Elise is an intense and highly intelligent young white Australian female living in Melbourne, Australia. She provides key insights in relation to self-care, based on her experiences juggling different activist roles and fostering cooperation among team members working on projects with long term, hard to define goals. She has completed studies in international development, has worked in international non-government organisations and academia, and has had a volunteering advocacy career running in parallel. She grew up in Darwin in the Northern Territory in a family and community that encouraged her to learn about the world and to explore issues and causes that were of interest. From a young age she travelled and was exposed to different cultures. She developed an awareness of societal challenges in relation to poverty, inequality, pollution, and environmental degradation, particularly for marginalised groups. She has undertaken a wide range of activism activities such as campaigning to increase the aid budget, encouraging divestment in companies that are not values driven, fundraising for various humanitarian crises, lobbying politicians, and cold calling individuals to influence their voting behaviour.

Iram

Iram is an Indian woman living in Dili, Timor-Leste. She is the co-founder of a social enterprise that aims to amplify the voices of marginalised groups, such as LGBTQIA+ people and victims of gender-based violence. Iram creates safe spaces for people to tell their stories, shares information on social media, attends Pride rallies and helps organise Rainbow festivals to reduce discrimination and violence for LGBTQIA+ communities in Timor-Leste.

Janelle

Janelle is a Caucasian New Zealander who is an activist for environmental issues. She has a professional background in science and law. She was previously heavily involved in animal welfare and climate change activism and in academia in the area of environmental law. Janelle was involved with coal blockades and protesting the opening of new coal mines. Through these experiences, Janelle learned a lot about group dynamics and how to manage her own energy levels whilst supporting the group to cooperate effectively. She currently works in an environment-focused role on the South Island of Aotearoa New Zealand.

Jenny

Jenny teaches high school English and English as an additional language (EAL) in her hometown of Melbourne, Victoria. She undertakes a wide variety of activities including being a foster carer, an online volunteer tutor for refugee students in detention in Indonesia, and an environmental, animal, and vegan activist. To better broadcast her activist messages, Jenny has put her hand up to run for parliament in state and federal elections. Growing up in an extended family of cattle farmers, her transition to becoming a vegan was a slow and considered process. As well as organising vegan community meals and cooking vegan treats for local events, Jenny currently has hundreds of subscribers on her Instagram site for people interested in making easy and quick vegan recipes.

Jody

Jody is a middle-aged Aboriginal Australian woman living in Kalgoorlie, Western Australia. She is quietly spoken, and humble, but also wise and comes across as having a strong sense of self. Although reflective and willing to take time to understand others' world views, she takes a stand against intolerance in her workplace. She works for the government operated health service as a consultant for the regional executive group. Her role involves providing cultural advice on policies, strategies, and programs that are implemented across the service. She advocates for changes that will benefit her community within the service. To complement her strong personal and professional experiences and bring a theoretical lens to her work, she has recently completed a Master's in Public Health. As the youngest of nine children and a mother of five children she also sees herself as an important role model and advocate for her family, always using ancestors and future generations as forces to fuel her motivation. Jody's experiences provide valuable insights for anyone navigating change across diverse cultures.

Khalil

Khalil is a Palestinian professional evaluator and doctoral candidate currently working in Switzerland and studying in Germany. His PhD topic surrounds the evaluation of equity and social justice. For more than 15 years, Khalil has been working as an internal and external evaluator in the international development sector. He has written on the topic of evaluators as activists and has a blog about agents of change. He is passionate about ensuring the voices of people of colour, women, and young people are included in his field, and he facilitates networks of professionals and administers forums on Twitter/X to help this become a reality.

Kim

Kim is an internationally awarded photographer and primary school teacher with New Forest Romany heritage. Living in Melbourne, Kim undertakes a variety of different and creative forms of activism. Among other things, she has been involved with issues around safeguarding land for wildlife, disability access and inclusion, refugee rights, and birdlife protection. She has used photography and other arts-based methods to create awareness, including through her highly popular lirralirra bird photography blogsite (tens of thousands of "likes"), lirralirra Facebook page (7200+ followers) and by starting an Ethical Bird Photography Facebook page. Kim has organised and marched in protests, held public meetings, lobbied politicians, written submissions for enquiries, interacted via social media, and spoken publicly on a range of topics.[7] She enjoys supplying images free of charge to not-for-profit animal welfare groups and activist groups as well as talking with environmental groups and camera clubs about ethical bird photography; she is pleased to see her passion about not disturbing, or shooting, birds gradually gaining awareness and acceptability.

Liz

Liz is a middle-aged Kenyan woman living in Nairobi and working for an international health organisation. She has strong family support for her activism and she draws upon her Christian values and beliefs to guide her actions. Her undergraduate degree was in education, but she knew from a young age that she wanted a career in international development. Her areas of interest include public health, child rights, community based environmental rehabilitation, and refugees. When she looks back on her career so far, her work supporting the economic empowerment of people living with HIV stands out as a highlight. However, she also reflects that a project she embarked upon with her sister that involved revegetating a small piece of land near her house was one of her most rewarding experiences. She uses that project not only to contribute to enhancing the environment where she lives, but as an activity to unwind and recharge. The way Liz articulates how her everyday activism makes her feel on an emotional level may help other activists consider the positive effects that activism can have on activists.

Marilee

Marilee has an energetic presence that fills up a room. Despite her strong character and self-confidence, she does not overpower the space; instead she shares her confidence with others in a way that empowers those around her. She grew up in a largely homogeneous Christian town in Alberta, Canada that provided her with a safe and quiet bubble. As she reached adulthood, she read books and engaged in discussions that questioned the narratives of her sheltered upbringing. Curious about the world and social justice, Marilee went off to university to study human rights and law. Her career has focused on tackling climate change and pursuing social justice. She currently holds an advocacy position in an international non-profit organisation. In her workplace and beyond, she has effectively advocated for women's equality and climate action.

Marion

Marion is a white Australian retiree who volunteers writing people's life story at a palliative care centre in Melbourne, Australia. The experiences of this wise and thoughtful woman with white hair and a large smile provide a long-term perspective on activism over the life course as she shares insights relating to her family, career, and retirement. She believes that social connections are key to influencing and changing people's lives. Throughout her life, Marion has attempted to be a positive example and make choices that align with her values. Her career began with teaching English as an additional language at a university and then she moved into international development. She worked with communities in Southeast Asia, the Pacific, and Africa. She also coordinated tertiary training opportunities for students at the Master's level of public health and evaluation. After she retired, she spent time walking El Camino in Spain before deciding where to place her time. In addition to volunteering with people in palliative care, she also volunteers with an organisation that offers support to navigate through the legal court system and donates money and time to organisations involved in political advocacy.

Matt

Matt is a software developer in Canada and, on the side, he is building an energy efficient house. In his spare time, Matt is a director with his community association board focusing on green initiatives including recycling, environmentally friendly transport options, and net-zero energy options for community infrastructure, as well as a community garden. Through careful collection of data, Matt can demonstrate how the local community association's diversion of waste from landfill is increasing year-on-year. This success has inspired a neighbouring community to follow in their recycling footsteps. Some other volunteer roles Matt has held include being a math and physics tutor and providing IT support for various community organisations.

MD Jamal

MD Jamal is a young Rohingya photographer who was born in a refugee camp in Cox's Bazar, Bangladesh. He has lived there for 27 years, since his family were forced to flee their home country of Myanmar in 1992. He works as freelance photojournalist documenting the life stories of his fellow refugees. He aims to show a global audience what life is like in the camps. MD Jamal has over 4000 followers on Twitter/X, including influential academics and policy makers in the international relations sector.

Mehreen

Mehreen is a first-generation Pakistani Canadian Ismaili Muslim with a passion for using technology in community development projects. She is always asking herself "how can we do the most good with technology?" Mehreen has a background in international development and her current role is Operations Manager at Red Iron Labs VR studio. She is the Prairie Regional Representative with the United Nations

Association in Canada and she uses this role to increase awareness about the UN and the Sustainable Development Goals. Mehreen is also involved with many local initiatives such as city clean-ups, community pantries, and general volunteering. She found satisfaction through her work with all the non-profit organisations with whom she has worked and she gets great joy from seeing her previous clients achieve their aspirations and flourish.

Mike

In alignment with his reflective, serious, and contemplative nature, Mike asked for the interview questions in advance. He provided written responses to the questions prior to discussing them in the interview. This allowed him time to contemplate and unpack the questions in more depth and resulted in elucidating an insightful perspective. Mike is a white Australian retiree living in Victoria who had a career as an evaluator in international development. He contributed to the development of programs focused on issues such as harm reduction, human trafficking, and economic development. He made recommendations that helped organisations avoid political interference, employ more local staff, and transform organisational culture to be more effective. He currently volunteers with a non-government organisation that offers support to people in palliative care. Over his lifetime, Mike says that if he achieved anything, it was facilitating and enabling the work of others.

Mosrat

Mosrat is a young female from Bangladesh who speaks quickly, directly, and succinctly. Her no nonsense efficient approach to talking is in alignment with the way Mosrat lives her life. She has a professional background in international development. She has recently been involved in a youth development program called Empower Youth for Work.[8] It focused on developing the leadership skills of young people from rural areas as well as practical job seeking skills. She has been able to keep in contact with some of the participants and feel pride knowing they have found employment through small businesses in their communities. Seeing some of the programs she has been involved with become self-sustainable makes her feel good. Mosrat provided pertinent advice about the importance of aligning work goals with an activist's inner passion. However, Mosrat's activism is not all about demanding ever higher levels of efficiency and effectiveness, she and her family also help a community group care for stray cats in their neighbourhood. The smile on her face as she shared the joy received from working with animals indicates the personal benefits she experiences.

MS Arif

MS Arif is a resilient Rohingya youth who serves as an academic researcher and a dedicated student of Political Science and Public Administration amidst the challenging backdrop of the Cox's Bazar camps in Bangladesh. He was born in Myanmar and endured the harrowing ordeals of a systematic genocide that drove his community to seek refuge in neighbouring Bangladesh in 2017. Despite grappling with severe limitations – minimal access to formal education, restricted freedom of movement, and scant prospects of returning home – MS Arif has displayed unwavering determination. He is tirelessly striving to secure opportunities for higher education, not just for himself but for the youth within his community. His personal commitment to amplifying their voices and labouring for their collective upliftment and the betterment of future generations is remarkable.

At present, MS Arif dedicates himself to advocacy work on behalf of his community and those in other refugee camps. Leveraging the power of social media, fostering youth cooperation, and engaging in direct action, he actively champions their cause. Through relentless efforts, including communication with humanitarian organisations and persistent advocacy, he has played a pivotal role in improving educational access within his camp. Furthermore, his initiatives have opened doors to educational opportunities for girls and women, thus transforming the lives of many within the refugee community.

Nafiz

Nafiz is an early middle-aged Bangladeshi man working for a company that facilitates ethical trading. He graduated from university with a Master's in Business, Economics, and Finance. However, it was early in his career when he decided to work for a non-government organisation because he witnessed a tragedy that occurred due to unsafe work practices. Volunteering in the cleanup was the catalyst for Nafiz deciding to shift his career to work for people who needed assistance. He is proud of this decision as, even though it was not necessarily in alignment with family expectations or career trajectory and the work can be physically demanding, he is infused with a passion to contribute to a better society.

This fuels his desire to do more and be better at what he does. His current role oversees the advocacy department that attempts to build alliances between companies, trade unions, and non-government organisations. This role brings him into areas such as environmental change resistance, labour rights, and gender-sensitive programming. Nafiz is proud of how the programs he has initiated have become self-sustaining and continue without his direct involvement.

Nicole

Nicole's energy levels are stratospheric. She is a vivacious, enthusiastic, and ambitious Aboriginal Larrakia woman in her mid-30s living and working in Darwin in the Northern Territory of Australia. Over the last ten years, Nicole has established herself as a driven leader. She is passionate about succession planning, intergenerational change, and empowering others. In addition to her full-time employment, she is an entrepreneur with her own business. She supports projects related to youth, women, and community development. Nicole is a fearless campaigner for improved communication and uses her extensive networks at a local, national, and international level to share positive narratives around First Nations peoples. Some examples include the work Nicole supports at the youth detention centre, developing a program for young entrepreneurs, and spearheading the organising committee for a national day of celebration for First Nations peoples in Darwin. Nicole adheres to a strict self-imposed self-care regimen that helps her to maintain her enthusiasm and energy over the long term.

Nigel

Nigel is a white British-Australian retiree living in a coastal town north of Sydney. He is inconspicuous, softly spoken, and talks in a humble way about his lifetime's worth of prolific and significant achievements. He was born in the United Kingdom and since he was a teenager he has been involved with political parties or environmental groups. Nigel obtained his Master's qualifications in planning from an American university then moved to Australia. Whilst working full time, Nigel volunteered with the Australian Privacy Foundation and the Consumer Federation of Australia advocating for consumers. He is currently involved with a local residents' association and holds an executive position on an environmental network which has had some success in advocating for environmentally sustainable development. He made a conscious decision to place a larger proportion of his efforts at the network level and successfully supported a range of small local groups to be more active. They are becoming self-sufficient and are able to contribute to the broader network, thereby increasing their impact.

Noraini

Noraini is an Indonesian entrepreneur with a gelato business. She also has paid employment in the banking sector and a regular voluntary role with an international non-profit organisation in migrant support and emergency services. She has found that her volunteer work provides her with a useful and satisfying way to use her time. She recalls highlights from her activism career where she has held positions that have been able to promote diversity. For example, when she was studying in the United States she organised a major event that incorporated the strengths of Indonesian and American culture.

Phyo Phae Thida

Phyo Phae Thida is a young PhD candidate, community organiser, and consultant for non-profit organisations. Her work and studies focus on promoting social change, diversity, and inclusion. She aims to raise awareness about the current situation in Myanmar through engaging with politicians, universities, civil society organisations, and the media. Although she has a gentle disposition, she is a beacon of determination and persistence. She grew up in a rural village in Myanmar where she was not expected to continue education past primary school. After researching a free school hundreds of kilometres away that would accept her, she begged her mother to let her go and at the tender age of 12 set out into the world alone to follow her dream of education. Since then, she has proactively sought out opportunities and excelled in everything she does, which brought her to a PhD scholarship at the University of Melbourne. Through her educational experiences on the topics of civic action and citizenship, Phyo Phae Thida began to speak out about injustices, particularly those affecting the people of Myanmar, including refugees residing in other nations. One of her highlights was providing peace leadership training to diverse young people and organising a panel discussion between people of different religions and ethnicities. Over 150 young people and representatives from the media attended. In the wake of Myanmar's military coup, she has also advocated to international communities including the Australian Government to take effective action against the military regime while supporting the people of Myanmar.

Putima

Putima is an Aboriginal and Torres Strait Islander grandmother from Larrakia Country, Darwin in the Northern Territory of Australia. She is young at heart and always has a smile on her face. Kindness and compassion radiate from her soul and anyone who talks to her can feel her empathetic energy. Putima has an extraordinary life story that has resulted in activism being embedded in her DNA. Her grandparents fought in the trade union movement and promoted the rights of Aboriginal and Torres Strait Islander Australians. After bearing witness to racism against her family in the aftermath of Cyclone Tracy in 1974, Putima subsequently stood up against racism aimed at the Vietnamese refugees who arrived by boat on the shores of Darwin and enrolled in her school. She has fought for many years to promote recognition of First Nations peoples in the Australian constitution. Putima also volunteers on the board of a local Aboriginal housing organisation, works full-time for a non-government organisation promoting health equity, co-cares for her elderly parents with her sister, and spends every spare minute with her precious grandchildren.

Ricki

Ricki is a very busy cat-loving woman who is on a staggering number of boards and committees as a chair, community advisor, ambassador, or member. Most of these organisations surround an inclusivity and advocacy agenda for LGBTQIA+ people with disability, health, Indigenous, and anti-violence focuses. She is also active in the academic and political worlds with roles in sociological groups and green politics. She has a Master's in Education from the University of Melbourne and is currently researching for her PhD. Ricki has a warm and generous nature and a compassionate heart with a fiery drive for social justice and inclusion. Her horrendous experiences of discrimination and violence for being transgender, Indigenous, autistic, and living with physical disability served to build her resolve to fight for others and cultivate a safer, more accepting world. Ricki's openness and willingness to take others on the journey is deeply inspiring; as is her penchant for head scarves and brightly coloured necklaces that say "look at me! I'm here and proud!"

Ronny

Ronny is a courageous creative man from West Papua living in Canberra. He is early middle aged, musically talented, and has been in Australia since he arrived to attend year 11 at high school. He has qualifications in social science and diplomacy and advocates for the self-determination of West Papua. Ronny has been involved in activism for the entire time he has lived in Australia. His family and friends back in West Papua need him and expect him to advocate on their behalf even though this places them in extreme danger. He uses every channel available to him to advocate for his people's independence. This includes supporting sporting programs, community radio shows, writing and performing music and protest songs, and coordinating music festivals. He is proud to be able to showcase the culture of West Papua on some of the biggest stages in Australia as he sees the invitations to perform as a sign of support for his cause and an opportunity to share his story.

Roy

Roy is a lawyer and author in his 70s who lives in a small town in Alberta, Canada. Over his lifespan, he has come to connect his Christian values with activism, which have morphed as he learns more about what he can do to have the highest impact. He recognises his privilege as an educated, white, male lawyer. He has sought to give back by focusing on his areas of interest, which are homelessness and deprivation among the First Nations Canadians in his local region and children living on the streets in Central America. He has travelled to Central America to work with these children but now predominantly provides and promotes financial support to organisations working on the frontline. In Canada, he mentors and supports First Nations people in an individual capacity.

Serena

Serena is a middle-aged woman from Papua New Guinea who does not necessarily consider herself to be an *activist*, but she was nominated by a colleague who said, "She is an amazing and inspiring woman who I know will not rest until she is able to educate every child in PNG." Serena is a lawyer by profession and when she was a law student, she and her friends founded a youth development organisation called The Voice Inc.[9] The organisation implements leadership and active citizenship programs with universities and high schools and helps young people find places to contribute to their communities outside their social group or their field of study. The organisation now has bases in all the major PNG universities to run extracurricular programs focused on personal development and on understanding legal rights. A number of other youth led, youth run organisations have also evolved as a result of The Voice Inc. Under her guidance, the organisation places emphasis on embedding reflection processes into everything they do. In addition to the organisational achievements, the inspiring part of Serena's story is how she pivoted from her original career despite the protestations of her family. She has since created something of which they are all now proud to be a part.

Susanne

Susanne is a white Australian living in Victoria. She has always worked for organisations that align with her values as an activist. She believes that she can increase her impact by incorporating activism into her paid employment. She is an advocate for the safety, support, and celebration of LGBTQIA+ people and works for an organisation that promotes inclusion and diversity. In addition to her work role, Susanne lives on five acres that has been converted from a paddock into a thriving garden. She appreciates any opportunity to pour energy into projects involving nature. When she reflects upon her career, she takes pride in knowing that she has trained educators, created referral pathways where there were none for trans and gender diverse people to seek support, and tried always to work for social justice and human rights.

Talia

Talia is a university student in her late teens. She is a first-generation English-Dutch Australian living in Victoria. At the age of eight she started a small business to fundraise for an organisation supporting children with autism after her cousin's diagnosis. She gained national publicity for her efforts at the time, and she continues her everyday actions that support recycling, repurposing, raising awareness, and minimising impact on the environment. She is aware of some individuals who have changed their attitudes because of her awareness raising and children who initiated their own responses after being inspired by her efforts.

Tamsin

Tamsin is a first-generation English and Canadian Australian mother of four young children, including 4-year-old twins, who lives in Victoria. The focus of her small business, Heaven and Earth Eco-Burial Products,[10] is to change the narratives around death practices. She wants to raise awareness about the damage caused by conventional

Western burial practices and move the industry towards more sustainable alternatives. Tamsin's choices throughout her life are also influenced by her values; she recycles, is a practicing yoga instructor, she lives a vegan lifestyle, walks instead of drives where possible, fundraises, promotes causes on social media, donates money, rehomes greyhounds, and actively promotes environmental and social justice agendas.

Tanya

Tanya is a white Australian living in Victoria on a property with a variety of rescue animals including horses. She is a vegan who works from home as a bookkeeper for clients including non-profit organisations who have a focus on rescuing and advocating for animals. As a person managing sometimes debilitating health issues, including depression, anxiety, and sleep disorders, she finds this flexible *behind the scenes* style of activism suits her own needs and that of the animals in her care. Tanya's motivation arises from knowing she is alleviating the burden of administration from other activists who are more drawn to the frontline. She takes pride in tracking the financial wellbeing of the organisations she works with and seeing them develop and become more established over time.

Thiha

Thiha came to Australia in 2012 as a refugee from Myanmar. During his early 20s in Myanmar he got involved in politics because he thought the disparity between rich and poor was unfair and there needed to be more equality and social justice. He was one of the student leaders involved in underground political activities against the authoritarian regime in Myanmar. He campaigned, planned protests, and secretly distributed pamphlets and handouts at a time when these activities were highly dangerous. For his efforts, he spent 18 years as a political prisoner in Myanmar. After becoming an Australian citizen in 2016, he has worked in community-based organisations that focus on community development. He is currently a case worker for asylum seekers with a non-profit organisation in Melbourne.

Vijeta

Vijeta is an Indian woman living in Melbourne and working and for a non-profit organisation as a social worker. She aspires to minimise her impact on society by being ethical and practicing minimalism. Professionally she finds safe homes for children in foster care. Within her organisation, she belongs to the LGBTQIA+ advisory group and promotes a culturally safe workplace through the wellbeing committee. She is also a wellbeing coach and is in the process of establishing a wellbeing coaching practice called Aikyam, which means *inner harmony*. She is diligent about maintaining her work, life, and family balance and enjoys spending time with her partner and creating happy moments for her dog, which in turn brings her great joy.

Notes

1 https://www.belisipng.org.pg/
2 https://heburials.com.au/
3 https://vickangas.org/
4 https://femilipng.org/
5 https://www.econetworkps.org/
6 https://www.angelslinkfoundation.com/
7 For example: https://www.ethicalnaturephotos.com/post/ethical-bird-photo graphy-by-kim-wormald
8 https://www.empoweryouthforwork.org/preparing-youth-for-the-future-of-work-bangladesh/
9 https://www.thevoicepng.org/
10 https://heburials.com.au/

1 Setting the scene

On a personal note

Throughout our lives, at times we have felt hopeless, overwhelmed, disheartened, frustrated, and discouraged about our efforts to save the planet. From a young age we have both had a desire to make positive change, but the road has not been easy. Family commitments, health issues, gender roles, financial obligations, employment duties, and other expectations, pressures, and demands have put us in some difficult circumstances and raised a myriad of ethical dilemmas.

We met at a conference about seven years before we started writing this book. However, there were some similarities in our life paths prior to meeting. Both growing up mainly in Australia in middle-income families, we were privileged enough to make choices about the course of our lives. We had experiences as children and young adults that prompted us to think about global issues, we were encouraged to think about career options beyond traditional gendered roles, and we were exposed to global travel and other cultures from an early age. We were aware that many others did not have those same opportunities. From when we were young, we felt a desire and a responsibility to use the resources we had at our disposal to work towards making the world more equitable and a more sustainable place to live.

A question about how to maximise our impact was a thread that continually ran through our respective journeys. As we struggled through periods of hopelessness and realised that we were never going to create the level of change that we desired, we never stopped searching for understanding. We wondered where we could best place our efforts to make the most difference and not become burnt out and disenfranchised.

At the conference where we met, we were separately presenting our research with people working in grassroots non-profit organisations.

DOI: 10.4324/9781003333982-1

Our doctoral topics intersected because we were both exploring how people like us, project level staff, were finding ways to demonstrate how they were making a difference. Since that meeting, we have shared our research findings and grown our friendship. At the core of this friendship is the consolation of finding someone to share the joys, frustrations, and confusions that occur as we attempt to navigate the complexities of a society that is ultimately structured in a way that makes change almost impossible.

The friendship has been synergistic. Instead of thinking we are battling alone, finding a fellow voyager with similar experiences and an inherent desire to interrogate their own reactions, thoughts, efforts, and positionality, gave us the courage to go even deeper. We know that we are not alone; there are millions like us, each connected to one another through various degrees of separation. We have read many guides aimed at leaders and people in senior positions. However, we have not found a book that talks to us, everyday people working for a better world, which provides ideas to frame and understand our actions and inactions and tools to help us measure, track, extend, and share our activist efforts.

Hence, as we start the next stage of our middle age lives, it is time to consolidate what we have learnt so far. This is important because we can see ourselves in so many other people. We are everyday people. Ordinary everyday people. However, we have been reading widely, researching with others in similar positions to us, thinking deeply, and constantly critically reflecting on our experiences. We have built careers and engaged in lifelong learning around making a positive difference in the world. We both work as non-profit practitioners as well as academics to combine theory, research, and practice. We are ready to share our accumulated wisdom as *pracademics* and continue the reflective process with a wider audience, incorporating the insights of 46 other everyday activists we interviewed for this book. While the ideas and techniques discussed throughout this book only scratch the surface, we invite others to share our synergistic critical reflective journey and unpack ways of maximising our collective power.

Purpose of this book

The overarching purpose of this book is to empower people who are not in leadership positions to understand how essential their everyday actions are for contributing to a better world. This book will help readers to maximise their capacity to influence change and overcome feelings of despondency that may cause inaction. It will explore some of the underlying motivations people have for taking action and discuss underlying philosophies related to finding purpose, meaning in life, and self-actualisation. A key part of the empowering aspect of this book is harnessing interpersonal skills to motivate others and constructively deal with people who disagree. It provides tools to incorporate research, theory, and evaluation to question, communicate, and inspire. It is our intention to support readers to set goals, track progress, celebrate wins, and practice self-care to enhance effectiveness and fulfilment.

This book is not going to list all the reasons why Planet Earth is at dire risk. The compendium of information to support this is easily accessible from many credible and scientifically validated sources (IPCC, 2021; Science and Security Board, 2022). We do, however, have a list of things that we believe are essential if humanity is to survive and thrive on Earth. We want to see an end to large scale intensive animal agriculture and deforestation, and a strong active commitment to managed rewilding, afforestation, and green urbanisation. We want technological solutions; a transformed energy base to renewables; decarbonisation; less waste; less production of commodities; and a break from acceleration, expansion, and never-ending growth. We want justice and equality; empowerment of marginalised groups; increased civic mindedness; and an ethic that puts kindness before profit, self-interest, and power. We want the support and participation of all people from every level and segment of society. We refer to many examples of initiatives that are working towards changes in these areas.

Throughout the book we draw upon our interviews with inspirational activists who are in low power *everyday* positions. We found people who are attempting to live up to Gandhi's (1999) suggestion that "If we could change ourselves, the tendencies in the world would also change" (p. 241). As detailed in the preliminary section on contributors, the types of people we interviewed included working and stay-at-home parents; Australian Aboriginal and Torres Strait Islander people (5); retirees; refugees and former refugees (6); disabled people (6); LGBTQIA+ people (4); students; researchers; freelancers; and small eco-business owners, among others. Interviewees included non-profit, for-profit, and government staff members and volunteers from Australia, Papua New Guinea, Bangladesh, France, United Kingdom, Fiji, Kenya, Nepal, South Sudan, Canada, Palestine, Myanmar (Bamar, Rakhine, and Rohingya), Brazil, China, Indonesia, India, Aotearoa, New Zealand, Pakistan, and the Philippines. The interviews with 46 everyday activists (31 female, 15 male) received ethics approval from Deakin University and were conducted throughout 2022. Interviewees were sourced through personal contacts and snowball sampling.

We also incorporate research that was conducted as part of our doctoral studies, with people who work in non-profit organisations. Our PhD research focused on how people in low power positions interrogated their work, questioned their contribution, and communicated their achievements. Similar to Arpad Szakolczai's (2021) approach where he suggests seeking information from a variety of sources, including unlikely guides, the empirical findings will be connected with existing research and theory from disciplines including sociology, philosophy, public health, and psychology. Quotations from fiction and nonfiction literature and cultural references provide examples to explore how characters found their passion from contributing service to others. We also draw upon our personal experiences. We share our individual and joint journeys for how we have found meaning in our lives and how we endeavour to make a small contribution to the betterment of the world.

This book focuses on self-reflective questions and ideas that can help people make a difference. We believe that the knowledge, research, capacity, and wisdom for how to make positive change already exists. We have not written this book to contribute new ideas and solutions – a plethora exist that just need to be implemented. We are aware, however, of the barriers that prevent application of great ideas in practice. As such, the focus is on providing theory and tools to support the application of changemaking ideas and solutions that are already in circulation, so that the likelihood of change increases. This chapter begins by introducing ourselves in more detail and articulating who might be interested in what we and our interviewees have to say. We define what we mean by everyday people and everyday actions and explain our understanding of how change can occur. We also share what we mean by effectiveness. We then illustrate with examples of why we think everyday people are so important before outlining the structure of this book.

Who are we to write this book?

Before we start, it is important for us to clearly and honestly locate ourselves along the spectrum of sentient privilege. We are highly educated, middle-aged, cisgendered, heterosexual white women living in a safe and developed country. We have health, good physical and mental abilities, secure accommodation, access to sources of income, and other privileges beyond the reach of many. Access to education and careers, made easier through many unearned opportunities, have given us a voice and platform to conduct our changemaking work, including equipping us with literacy and knowledge to write this book. By recognising our privilege, we are not saying that our lives are easy, or that we have not suffered intense loss, trauma, and lack of control over aspects of our lives governed by structures and actors with more power and privilege than us. We note that all sentient beings exist along this spectrum of privilege, and that we have more power and access to unearned opportunities than most human and non-human animals on the planet.

Seeing where we fit in the world is important for understanding our work as everyday activists and changemakers. Understanding our beliefs and values helps ground us in the *why* of our work. For example, as feminists, this means that we fight for equality and social justice; we recognise that socially constructed structures and demographics perpetuate the hierarchies that determine status, value, and power. We align our approach with author and activist Mikki Kendall (2020): "Feminism isn't just academic theory. It isn't a matter of saying the right words at the right time. Feminism is the work that you do, and the people you do it for who matter more than anything else" (p. 11). For us this means fighting for all humans to have agency over choices that affect their lives, including access to safe family planning, equal remuneration, equal distribution of caring roles and domestic tasks, informed consent, a non-discriminatory judicial system, safety, education, and healthcare. These rights apply to all humans, whether they live in a palace or an informal settlement, regardless of age, ability, ethnicity, gender, sexuality, and other social indicators that divide us. We acknowledge that our whiteness, class status, and able bodies have granted us access to an exclusionary and discriminatory system that prevents others from meaningful participation; we thereby take up our personal responsibility to reflect upon, receive feedback, and work to drive changes in behaviour and attitude across the system.

We aspire to be part of a movement that works to, as author and journalist Reni Eddo-Lodge (2018) states:

> liberate all people who have been economically, socially and culturally marginalised by an ideological system that has been designed for them to fail. That means disabled people, black people, trans people, women and non-binary people, LGB people and working-class people.
>
> (p. 180)

We make this point to highlight that we are all connected. By helping someone else, we help ourselves. This is not a one-way transfer of *us* helping *them*, it is working together to help each other through recognition that exploitation of one infects and affects us all. When we benefit from the exploitation of another being, negative consequences for us as individuals result. This point of view frames our work as activists and changemakers, not as altruistic and patronising *white saviours* on a mission to save the helpless masses (Althusser, 2014; Fanon, 1970) but as something that we are compelled to do to liberate ourselves. By working toward the freedom of others, and seeking to do no harm throughout our lives, we work toward our own freedom. This sentiment is most eloquently phrased by Aboriginal Elder, activist, and educator, Lilla Watson, and the Aboriginal activist groups that she was working with in the 1970s: "If you have come here to help me, you are wasting your time. But if you have come because your liberation is bound up with mine, then let us work together" (Lilla: International Women's Network, 2022, p. 1).

We do not deny, however, that a warm glow of altruism and other personal benefits can arise when contributing to something bigger than ourselves. For example, in one of Anton Chekhov's (1898/1987) plays a character described how his "heart swells with pride" when he cares for his neighbouring forest and plants a tree: "I feel as if I had some small share in improving the climate" (p. 30). Breaking the chains of oppression, working against exploitation of human, animal, plant, earth, sky, and sea, brings us closer to self-actualisation. While a volume of research exists to illustrate that social isolation is detrimental to health and wellbeing, the latest evidence from community psychology suggests that volunteers identifying with each other and focused on contributing to something outside themselves have a greater sense of belonging, build even stronger social support networks, and experience positive health benefits (Gray & Stevenson, 2020; Hoffman, 2018; Soren & Ryff, 2023; Yeung et al., 2017). Encapsulating this sentiment, Zadie Smith (2014) articulates the intense emotional response she felt when she contributed in a very small way during a community event that involved family and friends:

> I detected a hum of deep satisfaction at our many hands forming this useful human chain... Connected if only in gesture to an ancient line of practical women working in companionable silence in the service of their community. It's such a ludicrously tiny example of the collective action and yet clearly still so rare in my own life that even this minor instance of it struck me... Being with people, doing for people, it's going to bring you joy. Unexpectedly, it just feels better. It feels good to give your unique and prestigious selves a slip every now and then and confess your membership in this unwieldy collective called the human race.
>
> (p. 1)

This sense of purpose is often intrinsic but can be guided or supported by religious or spiritual beliefs. In this book we do not identify with any one religion, although many of our 46 interviewees without prompting noted their observation of Islam (6), Buddhism (3), Christianity (2), and Hinduism (1). We draw from several organised religions and philosophies including, but not limited to spiritualism, Christianity, Islam, Judaism, Quakerism, Taoism, Buddhism, and Bishnoi. We have worked for organisations that have had a religious doctrine as part of their underlying value system. We have worked alongside many colleagues during our careers whose religious convictions were the core driver of their work. While we acknowledge and respect people's right to follow their respective journeys, we have chosen a somewhat divergent and broader path. While we draw from and reference a variety of religious texts in this book, we also draw upon wisdom from philosophers, research from scientists, creatives, and insights from atheists and agnostics. These diverse sources have helped us find ways to harness our capacities to make a difference.

In essence, we wanted to write this book as a step along our own personal journeys. We are highly imperfect, hypocritical, and flawed, but we are working towards becoming our best, or better, selves. We are working towards being what Layla Saad (2020) terms "good ancestors" (p. 20). We recognised in ourselves that we had some knowledge and experience worth sharing, that could support others in their work. As everyday people who are striving to better the planet, we acknowledge that the entirety of our contributions is most definitely insignificant in the scheme of things. Our attempts pale in comparison to the work of many other everyday activists. However, part of the aim of this book is to reframe this level of contribution as a significant contribution to a greater whole. Our mini contributions, alongside yours, and those of millions of others, are building a groundswell of change.

Everyday people and everyday activism

Academics have discussed the concept of the *everyday* for centuries. Scholars such as Martin Heidegger (1927/2019) have framed everydayness (or *Alltäglichkeit* in German) as the realm of the mundane and ordinary. Historically, emphasis has been placed on the elite sphere where innovation and change were thought to occur. Thus, Max Weber (1978) outlines charismatic elites as the forces who have revolutionised society in contrast with the dullness of the everyday. This approach fails to recognise that the charismatic elites remembered in history books rarely worked alone and were usually grasped as token changemakers on account of their skin colour, station, and sex. The abolition of trans-Atlantic slavery is a good case in point. Often William Wilberforce, a wealthy and influential white male politician, first comes to mind in regards to this topic. However, the focus on one man obscures the efforts of hundreds of thousands of everyday activists who, at much greater personal cost than Wilberforce, revolted against their oppression. These people, and others who rose in solidarity with those enslaved, generated a tsunami of emancipation that rippled across nations from Haiti to Britain.

More recently, the *everyday* has become a focus in its own right. Scholars note the potential of the everyday as a space that can resist and challenge the status quo. Michel de Certeau (1984) and others (drawing from Michel Foucault and Pierre Bourdieu) explore the considerable agency enacted by everyday people. They recognise that it is everyday people who are continuously remaking and reclaiming the ideas, rules, and expectations forced upon them by dominant structures. This focus on the potential for everyday people to enact change is noted in many changemaking genres, including everyday peacebuilders (Mac Ginty, 2014; Ware et al., 2022) and everyday evaluators (Kelly & Rogers, 2022; Wadsworth, 2011), who are positively impacting their sphere of influence through small-scale, incremental efforts.

Evidence can also be found in popular culture. Amy Poehler's character, Leslie Knope, in the comedy series *Parks and Recreation* included a mention of the everyday as she described the joys of public service in a speech to new graduates from her old college:

> When we worked here together, we fought, scratched and clawed to make people's lives a tiny bit better. That's what public service is about: small, incremental change every day. Teddy Roosevelt once said, "Far and away the best prize that life has to offer is a chance to work hard at work worth doing."
>
> (Rosenberg, 2015, p. 1)

Another example of encouraging readers to celebrate and honour what could be considered mundane everyday work towards apparently unrealistic and unachievable goals, was captured by Julia Baird (2020) in her book *Phosphorescence:*

> [This is a story about people who] cared about something bigger than themselves, and who found their good intentions led them into a quagmire of boring meetings, and years of crushing inaction before things, finally, shifted or changed. It's also the story of those who tried, but never saw anything change, not in their lifetimes.

And it's about allowing ourselves to try, and honouring ourselves for caring, trying and giving a damn.

(p. 94)

When we talk about everyday people, we are talking about you. We are talking about everyone on the planet who is not in a position of high power. We are not talking about the small percentage of leaders, executives, high-level managers, board members of large organisations, or politicians with the power and influence to command obedience and generate top-down transformational change. Instead, we are talking about the millions of everyday people making positive differences in their corner of the world. These individuals, although relatively under-researched, have been found to influence other people, processes, and systems, even without formal authority (Bouquet & Birkinshaw, 2008; Gargiulo & Ertug, 2014; Hyde, 2018; Sozen, 2012). These are people of every age, ethnicity, and ability. Everyday people are acting to better the planet in support of those to whom they have a personal connection, such as migrants acting to help those in their home country or people raising money to fund research for a health issue that touched their family. People also mobilise for those beyond their personal sphere through acts such as boycotting companies that exploit workers, lobbying against cruel animal practices, choosing sustainable service providers, composting, recycling, and upcycling, avoiding animal products, creating backyard gardens that welcome creatures such as bees and birds, attending protests against refugee detention, and researching issues and solutions. People are mobilising in the slums of Nairobi, the refugee camps of Cox's Bazar, the embattled streets of Afghanistan, the high-rises of Hong Kong, the suburbs of Seattle, and the rural regions of Iceland.

We wrote this book for people who are not in positions where they can enact wide-reaching laws or make decisions that instantly affect the lives of millions. We wrote this book for the people Greta Thunberg is trying to inspire when she tweeted: "When enough people come together then change will come and we can achieve almost anything. So instead of looking for hope – start creating it" (Thunberg, 2021). Many people want to make positive changes in the world but can feel overwhelmed by their lack of influence. These people are probably not leaders; they may work as staff members in an organisation, run a small business, volunteer for a community group, be a stay-at-home parent, a retiree, a home-bound person, a researcher, or a student. They might work in an office as an administrator and spend weekends on the wetlands rescuing ducks and protesting the mass slaughter of native waterfowl during the duck shooting season. They might be incapacitated physically but write letters and run a social media campaign to prevent the loss of a biodiverse lake by opposing a new housing development. They might be a high school student who protests companies not paying their employees a liveable wage by sharing lists of naughty and nice companies and urging fellow students to boycott the naughty brands.

We see that most people are like us, in that they do not seek powerful leadership positions in society. We, the authors, do not want to be high level politicians, board members, or chief executive officers of influential organisations. We are content to be technical advisors and scholars, and recognise that many others also have no inclination for great power. However, this lack of desire for more power does not mean that we are willing to accept circumstances that exploit, cause suffering, and stifle the potential of others. An interviewee in Carol Gilligan's (1993) book expresses this sentiment perfectly:

> I have a very strong sense of being responsible to the world, that I can't just live for my enjoyment, but just the fact of being in the world gives me an obligation to do what I can to make the world a better place to live in, no matter how small a scale that may be on.
>
> (p. 21)

Thus, this book is written for us and for others like us, who want change and who are willing to act for it, but who have limited power to implement desired transformations. This book offers ideas and tools to support the good work – to help everyday activists track, extend, share, and celebrate small contributions toward saving the planet.

What we mean by saving the planet

We have thus far used the term *saving the planet*. We could have substituted this phrase with other equally cliched terms such as making the world a better place or contributing to improving the world. Lee (2016) suggests that "save humans" (p. 1) would assist with garnering the attention of more people, although we find this speciesist and reductionist. Other ways of articulating this sentiment include transformative change for betterment of people and the planet (Thomas & Luo, 2017; UNRISD, 2016), seeking sustainability (Paton, 2011), sustainable development (UNESCO, 2021), and improving planetary and human health (Myers & Frumkin, 2020). Mark McGillivray (2012, p. 24) "very loosely" describes development as "good change". Articulating the type and degree of change we want to see enables us to prioritise and make comparisons between intended and actual change. Peter Singer (2016) suggests that in order to live an ethical life we should enact effective altruism, which he defines as doing the "most good" possible, even if that means taking up a high-paid, less-fulfilling career to enable larger charitable donations to causes assessed as most worthy (p. iii).

We acknowledge that these types of phrases can mean a lot of things to different people. Change can be contentious. Some may be against change of any kind, positive or negative, preferring to hold on to the unrealistic notion that things can continue as they are indefinitely. Those who hold the power to define what is *good* and what is *improved*, and who can set the parameters for the intended change, may be in direct conflict with another group who defines change differently. The values of the group that are articulating the desired change may differ from those directly affected. Competing definitions of change and the implementation of various ways of understanding altruism and effective action will always occur. To be purposively provocative, a government may enact change without considering public opinion or deliberation. Or a mass murderer may annihilate the top capitalists rationalising that the elimination of humans is the most effective way to stop anthropocentric environmental destruction. However, while we acknowledge that the phrase *saving the planet* can mean different things to different audiences and is fraught with controversy, after considering the alternatives, for the purposes of this book we feel the phrase aptly serves our purpose. We are writing this book because we feel we do not have any choice but to move towards making changes that will benefit the planet, and hopefully save life on Earth, as quickly as possible. Hence, we have chosen *saving the planet* as an umbrella term because we are talking in the broadest possible sense, capturing our essential list above, to be as inclusive as possible.

We understand that each person has their reasons and interests for addressing a certain concern, and that their passion around those concerns is an important driver of change. We would like to use this book to remind our fellow humans who are interested in making positive change that all their efforts are worthy of celebration and have the potential to make change no matter how small. While we will share ways of questioning decisions about where to place efforts and how to make judgements about worthiness, we will not be offering our own value judgements as to where readers should focus their attention. We are also not going to debate the merits of one set of solutions over another. We understand there are ongoing passionate debates over cloth nappies versus biodegradable disposable nappies, hydrogen cars versus electric cars, use of hard plastics versus soft plastics, and a plethora of other debatable topics. We have chosen not to fall into the rabbit hole for these topics as we are happy to leave these discussions to others who have much more expertise than us.

Thereby, we are non-prescriptive about the issues throughout this book and discuss issues only to use them as examples to illustrate an approach or highlight a strategy in action. As new information comes to hand, and as the choice between alternatives becomes clearer, we will all need to re-evaluate our decisions. But until then, we encourage readers to learn from one another, be open to new ideas, support each other, and share information in a way that promotes meaningful conversations and positive ongoing interactions.

Change through cooperation

Like millions of others, we act for global betterment through small-scale interventions. Some of the things we personally do include avoiding consumption and use of animal products, working in and with non-profit organisations whose aims align with our values and goals, investing our superannuation in ethical funds, attending protests, contributing regular donations to a variety of non-profit organisations, signing petitions, dropping leaflets in letterboxes, knocking on doors, volunteering on stalls at festivals, boycotting exploiters, contributing to our local environmental protection and advocacy networks, minimising our use of fossil fuels, choosing renewable power, purchasing offsets, and lobbying influential leaders. We recycle and reuse in our homes, bring our cloth bags to the supermarket, carry our keep cups for takeaway coffee, and refrain from buying things we do not really need. We have conversations with friends and family about what more we could do. We foster and adopt dogs that need rehoming. We start neighbourhood initiatives and join local groups because of a desire to contribute to the wellbeing and social fabric of our communities.

Some of these actions we do because they are easy changes to make based on increasing awareness as information and new products come to hand (e.g., swapping single use plastic for beeswax wrap). Some are the result of personal connections to causes that have emotionally affected us into wanting to contribute (e.g., setting up a regular donation to a conservation group after an eco-adventure holiday). Some have occurred because of peer pressure and changing social norms that mean we would be outsiders in our social group if we did not do them (e.g., using paper straws or no straw at all). Others are because we have logically assessed where the points of leverage in our lives exist and used our power to select an ethical alternative (e.g., self-chosen superannuation fund). Some actions are because we have considered where our skill sets and education can best support others (e.g., providing research or writing assistance for our local community groups and seeking employment opportunities that match our values).

On a fundamental level we hold *cooperation* as a core tenet of our theory of how change happens. Morton Deutsch (2011) understood that people are interconnected, they influence each other, and when they have a common goal, they can be motivated to move towards achieving their ambitions. Cooperation and competition are the key components of the interdependence between people. We hold that when positive interdependent situations are established and cooperation results, then the trust, strong relationships, and effective ways of working together can result in synergistic achievements and improved psychological health (Deutsch, 2011; Johnson & Johnson, 2009). The importance of cooperation is central to Peter Kropotkin's evolutionary theory of *mutual aid* (Kropotkin, 1902/2012). He argues that humans and animals are unconsciously and instinctually driven towards solidarity and sociability and that this form of interdependence is a more important factor in evolution than mutual contest or competition. Kropotkin (1902/2012) states:

> It is the unconscious recognition of the force that is borrowed by each [being] from the practice of mutual aid; of the close dependency of everyone's happiness upon the happiness of all; and of the sense of justice, or equity, which brings the individual to consider the rights of every other individual as equal to his own.
>
> (p. 15)

More recently, themes of cooperation, mutual aid, and reciprocity are emerging in the literature that argue that altruism and cooperation on a group level are more effective than competition (Blaffer Hrdy, 2011; Servigne & Chapelle, 2022). Organisations that support people to work cooperatively to meet the needs of the group may incorporate the term *mutual aid* into their name or vision statement to distinguish their work from a traditional charity that has a one directional approach to support (Spade, 2020). However, it does not have to be directly transactional or immediately reciprocated, as it can simply mean "lending a hand" (Springer, 2020, p. 113). Springer (2020) suggests that the myriad of mundane yet pragmatic activities such as caring for a friend's children, carpooling, or taking a photo for someone you do not know, are examples of mutual aid being "the wellspring of all life on this planet" and considers it to be the "paramount element of our survival" (p. 113).

In relation to change, drawing from cooperation and mutual aid, we believe that change happens when many people do many different things. In environmental, political, and social struggles we need the quiet battlers, the protesters, the broadcasters, the social media warriors, the academics, and the extremists. We often criticise those who act differently from us, such as accusing quiet battlers for not doing enough or extremists for giving us a bad name. However, we believe we are collectively more effective when messages are communicated in many different ways. Exposing people to different forms of persuasive messaging increases the likelihood that they will take notice. Instead of worrying too much about activism perceived as problematic, this book supports readers to follow those who inspire them and create their own ways of acting to motivate others.

But perhaps you are wondering, how can one person make a difference when the system is stacked against them? Although we acknowledge the enormity of the systemic issues that prevent individuals from affecting change, we do not underestimate the potential for individuals to be the instigators of change. Our approach aligns with Amartya Sen (1999) who states:

> The freedom of agency that we individually have is inescapably qualified and constrained by the social, political and economic opportunities that are available to us. There is a deep complementarity between individual agency and social arrangements. It is important to give simultaneous recognition to the centrality of individual freedom *and* to the force of social influences on the extent and reach of individual freedom. To counter the problems that we face, we have to see individual freedom as a social commitment.
>
> (p. xii)

We interpret Sen's (1999) emphasis on "individual freedom as a social commitment" to mean that all individuals have a responsibility to work towards equity and equality for others. We believe that the freedom of an individual should not cause harm or exploitation of others. Hence, instead of being overwhelmed with how little one person can achieve, or focusing solely on our personal individual freedoms, we place our emphasis on Article 29 of the Declaration of Human Rights:

> Everyone has duties to the community in which alone the free and full development of [their] personality is possible... Securing due recognition and respect for the rights and freedoms of others and

of meeting the just requirements of morality, public order and the general welfare in a democratic society.

(United Nations, 1948, p. 1)

In essence, we believe that when enough people are aware and take action, then large corporations and governments will follow and enact the will of the majority. The approach that Margaret Mead adopted throughout her life provides us with a powerful affirmation: "Never doubt that a small group of thoughtful, committed citizens can change the world. Indeed it is the only thing that ever has" (The Institute for Intercultural Studies, 2022, p. 1). We find Barack Obama's (2009) work as a community organiser to be similarly inspiring:

> Because when you serve, it doesn't just improve your community, it makes you a part of your community. It breaks down walls. It fosters cooperation. And when that happens — when people set aside their differences, even for a moment, to work in common effort toward a common goal; when they struggle together, and sacrifice together, and learn from one another — then all things are possible.
>
> (p. 1)

Drawing strength from these sources and many other examples, we believe that humans have the collective capacity to cooperate and intentionally change the current system to save the planet.

What we mean by effectiveness

As authors, our point of difference is our careers as evaluators within non-profit organisations. For us, "Wanting to do good is not enough" (Theobald, 1999, p. 1). The *Encyclopedia of Evaluation* defines evaluation as a process of gathering and analysing evidence to make conclusions about "the state of affairs, value, merit, worth, significance, or quality of a program, product, person, policy, proposal, or plan" (Fournier, 2005, p. 140). Evaluation is a transdiscipline that developed from the social sciences to help make systematic judgements about the value of activities, initiatives, and social programs (Fournier, 2005; Scriven, 1991). Evaluation has the potential to help identify improvements, demonstrate achievements, and assist people to understand more about what they desire to do and what they are actually doing. Hence, when we use the term *effective* in this book we are referring to making judgements about the extent to which the activity delivers the intended outcomes or desired results (Davidson, 2005a). We attempt to think critically about our actions, interrogate our efforts, gather evidence, and systematically make informed assessments for the purposes of improving, communicating achievements, supporting decision making processes, identifying lessons learnt, and further contributing to the evidence base.

We have each spent years assessing the value, worth, merit, and significance of social programs, making evidence-informed recommendations for improving these programs, and tracking progression of change over time (Rogers et al., 2019b). We both got involved with evaluation because we saw the need for determining whether the ambitions of our respective organisations were being achieved. We have learnt about tools for measuring, sharing, learning from, and building on change through our work experience and critical intellectual investigations (Kelly, 2021; Kelly & Rogers, 2022; Rogers et al., 2019a, 2021). Seeing the connection between our personal activist goals and evaluation tools highlighted a potential opportunity to use the tools and approaches of the evaluation discipline at the personal level to measure, share, learn from, and build on our efforts to save the world.

Further, as highlighted in our opening paragraphs, our curiosity about how everyday people engage with evaluation to track, measure, and improve their efforts toward societal betterment drove us to undertake doctoral level research on topics focused on this phenomenon (Kelly, 2019, 2021; Rogers, 2021). Through our experiences and research, we have met many people who were not content with repeating things if they were not working or doing things just because they appeared to be working or felt good. Many workers, donors, community members, and clients want to know if they are funding, offering, and receiving a service or program that is going to deliver what was intended, bring benefit and not harm, and contribute to achieving an overarching goal. In addition, they also want to know what needs to be improved, how an initiative can be strengthened to achieve more, and how they can learn from and add value to the work of others.

Coming from a background in evaluation we understand how difficult it is to apply the logic of evaluation in practice (Kelly & Rogers, 2022). It can be difficult to find information and collect data. It can be hard to determine what the target audience will find meaningful. The constraints associated with time, capacity, and finances could mean that evaluation is not prioritised. The people involved may be reluctant to engage because they may feel anxious about reflecting on their work. The terminology associated with evaluation can be confusing and tools, approaches, and guidance can be pitched at a more complicated level than is necessary. For example, even the way we have used the term effectiveness may cause some evaluators to wonder why we have left out *appropriateness* or *efficiency*. Some types of evaluation focus on the process, others on the outcomes or impact, and others on economic aspects such as cost effectiveness and cost benefit analysis. Many guides that can navigate a path through these options and approaches to help find what suits your situation are available if more in-depth clarification is required (Better Evaluation, 2022; Davidson, 2005b).

However, in this book we are not interested in debating different definitions or methodologies. We encourage people not to get overwhelmed with the confusing aspects or think that they lack the technical knowledge or expertise to ask reflective questions about their work. Anyone can be an advocate for evaluation and explore meaningful and practical ways of judging effectiveness and determining whether their work is making a difference (Rogers et al., 2022; Wadsworth, 2011). We are interested in exploring ways of making sure that our efforts are useful, worthwhile, and valuable. From our experiences in the non-profit sector we understand the importance of being realistic and of acknowledging how identifying change can be extremely small, slow, hard to detect, and tricky to attribute (Kelly & Rogers, 2022). We also see the importance of everyday activists sharing their examples and the lessons they have learned with a broad audience. Sharing experiences, learnings, and achievements may make others more willing to join the journey and help people to feel part of something bigger.

Overview of the chapters

Each chapter of this book explores a different aspect of everyday activist practice. The first half of the book examines the inherent global structures and internal processes that depress, restrain, encourage, and motivate us. The second half is action-based and looks at how people can extend the impact of their work through measuring and broadcasting what they are doing to inspire others. Each chapter attempts to provide a more detailed response to the following questions:

- Why should individuals try when problems are so systemically entrenched? (Chapter Two)
- Why am I compelled to act? (Chapter Three)
- How can contributing help me work towards being my best self? (Chapter Four)
- How can I support and encourage others to get and stay involved? (Chapter Five)
- How do I deal with people who disagree? (Chapter Six)
- How will I know if I am making a difference? (Chapter Seven)

- How can I look after myself so I can contribute for longer? (Chapter Eight)
- How can I apply these strategies in my life? (Chapter Nine)

Chapter Two, "Facing a problematic world", shares how to stop feeling despondent and paralysed about huge problems and instead think about how each of us fits within the big picture. It draws upon systems thinking and sociology to elucidate the intersection between structure and agency, between the systems we live in and our personal goals, free will, and ability to enact change. The chapter unpacks the adage that ignorance is bliss, acknowledging the truth of this while uncovering how an existence of blinkered cognitive dissonance does ourselves and the world a disservice. Seeking to understand our place among the multilayered intersectional structures of sexism, homophobia, transphobia, classism, racism, ableism, ageism, and speciesism illuminates a web of privilege and disadvantage in which we are compelled to act, noting our complicity in systems of oppression. Rather than targeting everyday people as evil actors in these systems of oppression, this chapter draws on the concept of everyday agency to highlight the significant impact and social capital that can be achieved through pooling individual micro-actions. These micro-actions create a front of solidarity to transform local, national, and global systems.

The next chapter, "Exploring underlying motivations", discusses what drives activists into action. The chapter argues that it is worth taking the time to understand motivations, values, and beliefs so that you can articulate why you are doing what you do. Interviewees were motivated by a myriad of different drivers. Some included personal experiences, new knowledge, living in accordance with values, and a desire for justice. Others were motivated by duty, guilt, empathy, faith, family influences, and thoughts about legacy. This chapter references philosophers, psychologists, spiritual leaders, and creatives to understand how these motivations can assist with living a meaningful life. Determining whether our actions have a focus beyond the self and considering whether they are fulfilling, can provide insight into where we should be placing our efforts.

Chapter Four, "Finding fulfilment", focuses on how contributing as an activist can help with exploring personal identity and growing as an individual. It discusses the process of developing self-awareness and determining the degree to which activism contributes to becoming a self-actualised human being. Feeling inadequate, deflated, depleted, or like an imposter are common emotional responses among activists. The chapter argues these reactions, as well as positive energising effects, can be useful reflective tools for helping activists contribute in more constructive and effective ways for longer. Interrogating our emotional reactions can help find the balance between feeling good and continuing on the ongoing journey of personal development so that activists can become the best version of themselves.

Encouraging other people to get involved can be a powerful way of increasing impact. Chapter Five, "Cooperating with others", focuses on the synergistic effect of a diverse group of people working together. Effective teamwork can make challenging work less demanding and make it easier to find solutions. Sharing the joy and harnessing a wide range of skills can make problems seem surmountable. In contrast with thinking that we are in competition with others, we argue here that change requires cooperation – it is the platform upon which everyday activists must base their effective initiatives. The chapter draws upon a theory from social psychology to explain the key ingredients required for enhancing effective teamwork and dealing with negative group dynamics. Reflective questions challenge activists to think about how they establish shared goals, provide encouragement, develop social skills, be accountable, and reflect on group dynamics.

Chapter Six, "Dealing with people who disagree", acknowledges that dealing with deniers and defeatists can be emotionally draining. Although it might seem like a waste of energy to engage with people who have different points of view, this chapter argues that these courageous conversations are opportunities for growth and personal development. Researchers have identified effective strategies and the people interviewed for the book provide a spectrum of examples. Listening skills are a consistent theme throughout as only through understanding different perspectives and taking onboard new information can activists learn, revise and update plans, and modify and qualify their position. The chapter highlights the research-based interpersonal communication strategies that can help change attitudes and behaviours, particularly with people in our trusted social circles. Reflective questions prompt thinking about how to tailor interactions and how to increase the likelihood of achieving a mutually beneficial outcome.

Chapter Seven, "Everyday evaluation", emphasises the importance of drawing upon evidence and research findings before taking action to understand the underlying problem, articulate why change is needed, and guide the development of actions. We discuss the importance of reflecting, tracking, and sharing activist efforts. Evaluation is highlighted as a tool that can assist in measuring and articulating change. We provide actionable guidance and link to tools that can support everyday activists to consider the big picture and articulate how activities will contribute to higher level outcomes and goals. The concept of knowledge translation is introduced as a means to make sense of and disseminate information, thus equipping *sharing* as an act of activism in itself. The chapter concludes with practical ways of incorporating frameworks and models to help activists to think beyond just counting how many things are done and work towards articulating quality and impact to enable determination of effectiveness.

Chapter Eight, "Self-care", focuses on psychological strategies such as maintaining balance, avoiding burnout, finding joy, and being grounded in the present. This chapter draws from disciplines, such as social work and emergency services, where professionals face significant vicarious trauma and are likely to suffer burnout if their self-care is not prioritised. The literature is augmented through the empirical findings whereby people interviewed for this book share their techniques for finding balance between their changemaking work and the other aspects of their lives.

The conclusion brings together the key strategies from each chapter into a practical guide for application in the real world. Tools and examples for overcoming inertia, identifying motivations, finding fulfilment and harnessing skills, sharing work, engaging with others, questioning effectiveness, and practicing self-care are presented to provide inspiration and guidance in an accessible format. We conclude the book by presenting our suggestions for navigating future issues. We share examples of how adopting a learning mindset, listening carefully, continually developing self-awareness, and slowing down can help everyday activists to be more effective over the long term. We advocate for the uptake of these suggestions so that ultimately we can collectively upscale our efforts and enhance our contributions.

Big picture vision

We started this chapter talking about why the topic of everyday people saving the planet is so important to us. As stated above, we do not have *the answers*, but we have critically reflected on our journey so far. By documenting our engagement with the literature and our discussions with other everyday activists, this book offers strategies to help readers leverage their power for maximum effect. We hope this book provides reinvigoration through sometimes overwhelming feelings of helplessness and frustration. While we acknowledge that we still have these feelings, through the process of writing this book we are now more mindful of where and how they originate. We have the strategies to use these emotional reactions to fuel more action and promote more positive change. We hope that this book will assist readers to channel their emotional responses towards maximising their collective power to affect positive change. We want everyday activists to recognise the significant role that they have in dismantling systems of oppression and reimagining the way the world works. Together, we can work towards making the world a better place whilst developing our own sense of self – by helping others, we help ourselves. We hope that this book will help encourage everyday people to keep going and keep battling. Together our micro-efforts will connect, amplify, and transform.

References

Althusser, L. (2014). *On the reproduction of capitalism: Ideology and ideological state apparatuses.* Verso.

Baird, J. (2020). *Phosphorescence: On awe, wonder & things that sustain you when the world goes dark.* HarperCollins.

Better Evaluation. (2022). *Manager's guide to evaluation.* https://www.betterevaluation.org/en/managers_guide

Blaffer Hrdy, S. (2011). *Mothers and others: The evolutionary origins of mutual understanding.* Harvard University Press.

Bouquet, C., & Birkinshaw, J. (2008). Managing power in the multinational corporation: How low-power actors gain influence. *Journal of Management,* 34(3), 477–508. doi: https://doi.org/10.1177/0149206308316062

de Certeau, M. (1984). *The practice of everyday life.* University of California Press.

Chekhov, A. (1987). *Uncle Vanya: Scenes from country life in four acts.* Heinemann (Original work published 1898).

Davidson, E. J. (2005a). Effectiveness. In S. Mathison (Ed.), *Encyclopedia of Evaluation* (p. 122). Sage Publications. doi: https://doi.org/10.4135/9781412950558

Davidson, E. J. (2005b). *Evaluation methodology basics: The nuts and bolts of sound evaluation.* Sage Publications. doi: https://doi.org/10.4135/9781452230115

Deutsch, M. (2011). Cooperation and competition. In P. T. Coleman (Ed.), *Conflict, interdependence, and justice: The intellectual legacy of Morton Deutsch* (pp. 23–40). Springer. doi: https://doi.org/10.1007/978-1-4419-9994-8_2

Eddo-Lodge, R. (2018). *Why I'm no longer talking to white people about race* (2nd edn.). Bloomsbury Publishing.

Fanon, F. (1970). *Black skin, white masks.* Paladin.

Fournier, D. M. (2005). Evaluation. In S. Mathison (Ed.), *Encyclopedia of evaluation* (p. 140). Sage Publications. doi: https://doi.org/10.4135/9781412950558

Gandhi, M. (1999). *The collected works of Mahatma Gandhi.* Publications Division Government of India.

Gargiulo, M., & Ertug, G. (2014). The power of the weak. *Research in the Sociology of Organizations,* 40, 179–198. doi: https://doi.org/10.1108/S0733-558X(2014)0000040009

Gilligan, C. (1993). *In a different voice: Psychological theory and women's development.* Harvard University Press.

Gray, D., & Stevenson, C. (2020). How can 'we' help? Exploring the role of shared social identity in the experiences and benefits of volunteering. *Journal of Community & Applied Social Psychology*, 30(4), 341–353. doi: https://doi.org/10.1002/CASP.2448

Heidegger, M. (2019). *Being and time*. Martino Fine Books. (Original work published 1927).

Hoffman, J. A. (2018). Community gardening, volunteerism and personal happiness: "Digging in" to green space environments for improved health. *Psychiatry, Depression & Anxiety*, 4, 1–7. doi: https://doi.org/10.24966/PDA-0150/100015

Hyde, C. A. (2018). Leading from below: Low-power actors as organizational change agents. *Human Service Organizations: Management, Leadership & Governance*, 42(1), 53–67. doi: https://doi.org/10.1080/23303131.2017.1360229

IPCC. (2021). *Climate change 2021: The physical science basis - Contribution of working group I to the sixth assessment report of the intergovernmental panel on climate change*. Cambridge University Press. https://www.ipcc.ch/report/ar6/wg1/#FullReport

Johnson, D. W., & Johnson, R. (2009). An educational psychology success story: Social interdependence theory and cooperative learning. *Educational Researcher*, 38(5), 365–379. doi: https://doi.org/10.3102/0013189X09339057

Kelly, L. (2019). *What's the point? Program evaluation in small community development NGOs*. Deakin University, Geelong.

Kelly, L. (2021). *Evaluation in small development non-profits: Deadends, victories, and alternative routes*. Palgrave Macmillan. doi: https://doi.org/10.1007/978-3-030-58979-0

Kelly, L., & Rogers, A. (2022). *Internal evaluation in non-profit organisations: Practitioner perspectives on theory, research and practice*. Routledge.

Kendall, M. (2020). *Hood feminism: Notes from the women white feminists forgot*. Bloomsbury.

Kropotkin, P. (2012). *Mutual aid: A factor of evolution*. Dover Publications (Original published in 1902).

Lee, B. Y. (2016). "Save the planet" really should be "save humans." *Forbes*. https://www.forbes.com/sites/brucelee/2016/11/13/why-you-shouldnt-tell-donald-trump-to-save-the-planet/?sh=6ad1be85e2ac.

Lilla: International Women's Network. (2022). *Lilla: International Women's Network*. https://lillanetwork.wordpress.com/about/

Mac Ginty, R. (2014). Everyday peace: Bottom-up and local agency in conflict-affected societies. *Security Dialogue*, 45(6), 548–564. doi: https://doi.org/10.1177/0967010614550899

McGillivray, M. (2012). What is development? In D. Kingsbury, J. McKay, J. Hunt, M. McGillivray, & M. Clarke (Eds.), *International development: Issues and challenges* (2nd edn., pp. 23–52). Palgrave Macmillan.

Myers, S. S., & Frumkin, H. (2020). *Planetary health: Protecting nature to protect ourselves*. Island Press.

Obama, B. (2009). *Barack Obama at Notre Dame: Commencement Address Transcript.* https://time.com/4336922/obama-commencement-speech-transcript-notre-dame/

Paton, G. J. (2011). *Seeking sustainability: On the prospect of an ecological liberalism.* Routledge.

Rogers, A. (2021). *Competitive champions versus cooperative advocates: Understanding evaluation advocates in Australian non-profit organisations.* Doctoral thesis, Centre for Program Evaluation, University of Melbourne, https://minerva-access.unimelb.edu.au/handle/11343/288857

Rogers, A., Gullickson, A. M., King, J. A., & McKinley, E. (2022). Competitive champions versus cooperative advocates: Understanding advocates for evaluation. *Journal of MultiDisciplinary Evaluation,* 18(42), 73–91.

Rogers, A., Kelly, L., & McCoy, A. (2019a). Evaluation literacy: Perspectives of internal evaluators in non-government organizations. *Canadian Journal of Program Evaluation,* 34(1), 1–20. doi: https://doi.org/10.3138/cjpe.42190

Rogers, A., Kelly, L., & McCoy, A. (2019b). Pathways to becoming an internal evaluator: Perspectives from the Australian non-government sector. *Evaluation and Program Planning,* 74, 102–109. doi: https://doi.org/10.1016/j.evalp rogplan.2019.01.007

Rogers, A., Kelly, L., & McCoy, A. (2021). Using social psychology to constructively involve colleagues in internal evaluation. *American Journal of Evaluation,* 42(4), 541–558. doi: https://doi.org/10.1177/1098214020959465

Rosenberg, A. (2015). *What 'Parks and Recreation' taught me about life. The Washington Post.* https://www.washingtonpost.com/news/act-four/wp/2015/02/25/what-parks-and-recreation-taught-me-about-life/

Saad, L. (2020). *Me and white supremacy: How to recognise your privilege, combat racism and change the world.* Quercus.

Science and Security Board. (2022). *2022 Doomsday clock statement.* Bulletin of the Atomic Scientists. https://thebulletin.org/doomsday-clock/current-time/

Scriven, M. (1991). *Evaluation thesaurus* (4th edn.). Sage Publications.

Sen, A. (1999). *Development as freedom.* Oxford University Press.

Servigne, P., & Chapelle, G. (2022). *Mutual aid: The other law of the jungle.* John Wiley & Sons.

Singer, P. (2016). *The most good you can do: How effective altruism is changing ideas about living ethically.* Text Publishing.

Smith, Z. (2014). *Don't let your fellow humans be alien to you: Commencement speech - Many Hands, The New School.* Speakola: Commencement and Graduation. https://speakola.com/grad/zadie-smith-many-hands-2014

Soren, A., & Ryff, C. D. (2023). Meaningful work, well-being, and health: Enacting a eudaimonic vision. *International Journal of Environmental Research and Public Health,* 20 (16), 6570. doi: https://doi.org/10.3390/ijerp h20166570

Sozen, H. C. (2012). Social networks and power in organizations: A research on the roles and positions of the junior level secretaries in an organizational network. *Personnel Review*, 41(4), 487–512. doi: https://doi.org/10.1108/00483481211229393/FULL/XML

Spade, D. (2020). Solidarity not charity: Mutual aid for mobilization and survival. *Social Text*, 38(1), 131–151. doi: https://doi.org/10.1215/01642472-7971139

Springer, S. (2020). Caring geographies: The COVID-19 interregnum and a return to mutual aid. *Dialogues in Human Geography*, 10(2), 112–115. doi: https://doi.org/10.1177/2043820620931277

Szakolczai, A. (2021). *Post-truth society: A political anthropology of trickster logic*. Routledge. doi: https://doi.org/10.4324/9781003225553

The Institute for Intercultural Studies. (2022). *Frequently asked questions about Mead and Bateson*. http://www.interculturalstudies.org/faq.html

Theobald, R. (1999). *We do have future choices: Strategies for fundamentally changing the 21st century.* Southern Cross University Press.

Thomas, V., & Luo, X. (2017). *Multilateral banks and the development process: Vital links in the results chain*. Routledge. doi: https://doi.org/10.4324/9781315124865

Thunberg, G. (2021). *Tweet by Greta Thunberg on Twitter*. https://twitter.com/gretathunberg?s=11

UNESCO. (2021). *Sustainable development*. https://en.unesco.org/themes/education-sustainable-development/what-is-esd/sd

United Nations. (1948). *Universal declaration of human rights*. https://www.un.org/en/about-us/universal-declaration-of-human-rights

UNRISD. (2016). *Policy Innovations for Transformative Change: UNRISD Flagship Report*. United Nations Research Institute for Social Development.

Wadsworth, Y. (2011). *Everyday evaluation on the run* (3rd edn.). Allen & Unwin.

Ware, A., Ware, V.-A., & Kelly, L. (2022). Strengthening everyday peace formation after ethnic cleansing: operationalising a framework in Myanmar's Rohingya conflict. *Third World Quarterly*. Doi: https://doi.org/10.1080/01436597.2021.2022469

Weber, M. (1978). *Economy and society: An outline of interpretive sociology*. University of California Press. (Original work published 1922).

Yeung, J. W. K., Zhang, Z., & Kim, T. Y. (2017). Volunteering and health benefits in general adults: Cumulative effects and forms. *BMC Public Health*, 18(1), 1–8. doi: https://doi.org/10.1186/S12889-017-4561-8/TABLES/4

2 Facing a problematic world

This chapter is about recognising that we are all part of the problem and part of the solution. We begin by drawing upon sociology to help understand the concept of power and how global problems are so entrenched in the systems and structures in which we live. Faced with this evidence, it is no wonder that we can feel despondent and overwhelmed in a manner that can prevent us from acting. Feeling apathetic or complacent, becoming critical, blaming the victim, or getting defensive are all possible unhelpful ways of coping. However, there are alternative responses where everyday people can instigate change.

Everyday people have agency and the ability to create change. This chapter uses frameworks drawn from everyday peace and everyday resistance. It looks at the relationship between social structures and individual actions and the potential for collectiveness and cooperation to unite people behind a common goal. We share how individuals can act for change on a global level. We conclude the chapter by sharing practical strategies for finding the strength to act in the face of real and significant barriers to change.

The problematic world

Sociological theories that explain how the world functions centre on the concept of power. This power is exercised through the neoliberal model based on free market capitalism and deregulation that values individualism, consumerism, and competition, and sees the Earth as something to exploit and control. These theories propose that the world is a complex system of structures like gender, race, and class, where a small group of powerful people have dominance over all other entities.

While colonisation and plunder have occurred for centuries, the world changed exponentially over the past few hundred years. The industrial revolution set off a chain of engineering and scientific

DOI: 10.4324/9781003333982-2

inventions that transformed every aspect of society in the Global North. Peasants flocked to cities, food production increased, health outcomes improved, more children survived past childhood, knowledge grew, inventions proliferated, and education systems developed. At the same time, this growth necessitated an accelerating demand for natural and human resources. Unable to meet demand by plundering the Global North, nations leading the industrial revolution raced to seek natural and human capital elsewhere. They were on a mission to pillage, steal, and overpower unsuspecting regions across the Global South, from Asia and Australasia to the Americas and Africa.

In carving up the Global South, power was held by a small percentage of upper-class white men. It is important to make this distinction as it was not the poor, or women, or people of colour who created the hierarchies and institutions of today's world. This is important because modern structures have been shaped by these rich white men who saw themselves as the norm against which all others should be measured. Their power has reverberated throughout the centuries, constantly reinforced, and their mantle accepted by the next generation of rich white men. Although this power was centralised in the hands of a few, enough of it trickled down to those below to lift our living standards and keep those of us most similar to those rich white men on side, willing to reinforce systems of control that provide us with adequate shelter, healthcare, education, civic infrastructure, and a feeling of superiority.

This feeling of superiority creates layers of dominion, whereby humans take control over the environment and non-human animals. Within humankind, these layers are systemically etched into our society. The layers of power are so embedded that sometimes we can barely see them, especially when they are to our advantage and disadvantage others outside our purview. When we do spend the time understanding how the issues are connected and entwined within systemic layers of oppression, we can become overwhelmed. Our lack of power against these structures is seemingly insurmountable. In the face of such an immense and largely intractable system, we can be paralysed into inaction.

Many of the activists we spoke to were aware of the power dynamics underlying the surface level injustices. Ricki illuminated the systemic influences that "reinforce power and control over others." She interweaved her hands to depict "a schema that connects us to the big picture of why we are here, what are our resources, why do we allocate to certain groups, and why do certain groups always seem to be disadvantaged." Others highlighted the macro forces that actively work against them. Our fellow activists in places like Myanmar, South

Sudan, and Bangladesh noted the government opposition to their work and how they need to diplomatically reframe what they are doing or attempt to continue under the radar. Matt raised that these resistant forces are at play globally, referring to the election of governments in Western nations "that are fairly climate change denying" and the "disheartening" barriers that imposes. Alex explained that sometimes it feels pointless to act if the government is not interested in addressing the issues at the systemic level: "there are all these people like me doing these little things that all add up, but… it won't make a difference if the politicians don't represent that."

When asked if they ever feel overwhelmed, disillusioned, or demotivated in their activism, our fellow everyday activists answered affirmatively. Adrian stated emphatically, "Oh yeah. Heck yeah. Heck yeah." Kim outlined that "most people in it feel disillusioned from time to time because it's hard work." Ekawati noted that she has "complained about these feelings" but is "nowhere close to an answer." Tamsin highlighted the frustration of playing the long game: "Like with issues that are going on for so long and every time you think you're getting somewhere then it turns out that you're not. Or when people don't listen or don't even try something (e.g., vegan food). I find ignorance really depressing, where it doesn't matter what you say, they're not going to hear you." Similarly, Alyssa noted how exhausting and frustrating it is when you "try and try again and nothing changes."

There were other reasons why interviewees sometimes felt inertia. Some interviewees recognised that their impact is likely infinitesimal and slow. Janelle was overcome by a feeling of helplessness after having invested her heart and soul into environmental activism in the Antipodes. She was in China in a traffic jam with "eight lanes this way and eight lanes that way and it was just full of cars and full of people… There were probably more people in that traffic jam than there are living in New Zealand." At the time she felt "exceptionally dejected about the scale of things in the rest of the world compared to the tiny little things that we'd been doing and achieving in Australia." Marilee also "feel[s] quite nihilist about it sometimes" and feels daunted by "the enormity of the task and the likelihood of success". Some interviewees noted that "Change takes forever, and nothing comes fast" (Clariana), and that "change is extremely slow and can't be felt all the time" (Iram).

As well as the often painfully slow pace of change, Alyssa noted how dealing with difficult personalities can be intensely draining and demotivating. After being trolled, doxed, and slandered, including from *allies* within the movement, there were times when she wondered "if maybe I should just walk away". This sentiment was raised by several activists who mentioned the heartrending pain of being slandered or otherwise maligned by those ostensibly fighting the same fight. They mentioned how much of themselves they poured into their respective causes and how much more that made the betrayal of seemingly likeminded people sting their souls. Alyssa drew it back to suggest that many of these people "just linger and they're not effective anymore... they get tired, and a bit crazy. They've just gone defunct." This raises the issue of when to stop and take a break, which will be discussed later in Chapter Eight on self-care. At the same time, it raises another concern voiced by our activists whereby they feel like "they're not enough" (Aruna).

Many of the interviewees highlighted that not knowing the impact of their work contributed to feelings of overwhelming disillusionment. Roy, an activist in his 70s, explained that he does not really know if his work makes a lasting difference: "I feel discouraged more than encouraged. I've done this stuff for years and you just kind of never really know." Similarly, teenager Talia clarified that:

> Sometimes not being able to see the impact of your contribution can make you feel demotivated. I think the media very much talks about all the issues and problems around climate change and it can feel really overwhelming and deflating like I've done so much yet it's not enough. I think it's really hard with really big problems and not being able to see the repercussions of your good actions. I think to overcome them, it can be really hard and you get stuck in a negative cycle, and it's reminding yourself that you're playing your part and even though you don't have control over other people's actions, you're in control of your actions and you can make them good ones and inspire someone else.

Chapter Three provides an in-depth exploration of the underlying motivations of everyday activists to examine how people pull themselves out of a state of inertia into action. The next part of this chapter focuses on unhelpful coping strategies so that readers can avoid succumbing to these negative thought patterns. It draws upon sociological theory to provide some uplifting and empowering ways forward.

Common mechanisms for coping with the problematic world

Many of our fellow everyday activists raised how common it was for people in their networks to display general apathy. Alyssa highlighted that dealing with apathy was the hardest part about being an activist. She described her activism, which is around native animal torture, as being "very hard on the heart" in general. "Not just dealing with cruelty, but also dealing with apathy, which is even worse." Tamsin noted that "So many people aren't interested in changing at all, and they don't care. People get defensive, because deep down they think they're doing something wrong. Maybe then you should change it?"

When it feels like there are so many problems in the world, and it is all too much, our minds have ways of helping us deal with our thoughts. Apathy can, in part, be explained as a form of self-protection. However, these thoughts can leave us trapped in a false world that is disconnected from fact. Psychologist Leon Festinger (1957) coined the term cognitive dissonance as a way of explaining how people engage in self-delusion to help them cope with their opposing or inconsistent ideologies.

> Cognitive dissonance is a state of tension that occurs whenever a person holds two cognitions (ideas, attitudes, beliefs, opinions) that are psychologically inconsistent... Dissonance produces mental discomfort, ranging from minor pangs to deep anguish; people don't rest easy until they find a way to reduce it.
>
> (Tavris & Aronson, 2007, p. 13)

Essentially, cognitive dissonance is a psychological tool that allows us to justify doing nothing about things that we know are wrong. It can provide legitimacy to follow an easier route in life and ignore those small voices in our minds that tell us we should act otherwise. Thus, confronting and deconstructing these mental structures can be self-liberating.

Complacency can also be driven by fear, whether acknowledged or internalised. Alicia discussed how she saw terrible acts of violence occur from her office window in Papua New Guinea. Earlier in her career, the frequency and commonality of these acts lulled her into a false sense of normalcy and made her feel unable to change the situation. Addressing the violence would have put her at risk; remaining quiet and unseeing was self-protective. Noraini mentioned that this same dynamic plays out at all different levels, not just life-threatening ones. Often people do not speak out or get involved because they want to be liked and not labelled as overly sensitive or a troublemaker.

Monetary arguments can help rationalise contributions to global problems. Economic reasons can convince people that something unethical or planet destroying is acceptable in much the same way as fear and pleasure can convince people that staying out of an unjust situation or treating animals cruelly is acceptable. Adrian highlighted that big corporations are "all very good on the corporate social responsibility, particularly in terms of environment and phasing out coal and fossil fuels, but when the dollar speaks, they sign the contract." He noted that "They're not walking their talk" because money motivates enough to outweigh those nagging feelings of something not sitting right.

Defensiveness is a common reaction people use to reduce dissonance. This often manifests as a need to deride those who are trying to do good. Some people may soothe their dissonance by criticising *do-gooders* as acting for their own advantage. Phyo Pyae Thida and Alyssa commented on the devasting effect personal attacks can have on activists' mental health. Both women invest their heart and souls into their respective volunteer causes, at zero gain for themselves and put them at risk of physical and verbal attack. Both have experienced accusations of seeking the limelight and using activism to climb some invisible career ladder. Despite the crushing pain, the importance of their causes compelled them to continue pushing ahead.

A misconception exists that activism is a privilege that is exercised by people who essentially have the time and money to invest in pushing for positive change. This assumption is fundamentally untrue globally and throughout history. Many of the poorest and most oppressed groups have fought for change despite extreme personal and group risk. A clear example in our cohort of everyday activists are the young Rohingya men we spoke to who live in abject conditions in what is currently the world's largest refugee camp. These activists, alongside thousands of others in the camps, persistently broadcast their situation, reminding the world of their plight through photojournalism and poetry, beseeching the world not to forget them. Others in our cohort, such as those advocating for West Papua and in South Sudan, face death threats and the real risk of violence for their activism. So, activism is not the arena of those rich in time, money, and other resources. In fact, Iram suggested that the opposite is true commenting that "What discouraged me most about the privileged human beings who believe they are entitled to all the goodness in life is that they have nothing great to offer to others."

Apathy, cognitive dissonance, inaction of others, excuses, and criticisms can be hard to take. Silence enables and replicates distorted structures that manipulate and justify a blinkered view of the world. Our fellow activists sometimes felt despairingly about humanity and lost hope for a better future. However, while demotivation and disillusionment can push activists to stop trying, and can be a sign that they need to take a break or rethink their self-care strategies (see Chapter Eight), they often find reinvigoration or a tenacity to keep going buried in their core values and principles. Instead of feeling "burnt out and overwhelmed because Mumbai city presents you with a lifetime worth of work" (Dia), or worrying that "it's hard against the enormity of things to feel like you're making much of a difference" (Elise), we can take comfort in recognising that we are part of a larger movement to disrupt and deconstruct these systems and the myriad symptoms they cause. As such, we are not working alone, there are millions of us working together. Then, Aruna's feeling: "I can't do everything that needs to be done myself and I don't have the level of influence I wish I could have" and Nafiz's recognition that: "I don't have that much of a leadership role to change any major things" become less of a barrier, because we are one of many.

What we can do about the problematic world

There are numerous examples from history where micro-efforts by individuals have led to significant global transformations. The Civil Rights Movement of the 1950s and 1960s in the United States was driven by the collective efforts of those who fought against racial segregation and discrimination. Ordinary people took individual stands that sparked a larger movement for civil rights. Their micro-efforts, along with the activism of many others, ultimately led to the passage of key legislation, including the Civil Rights Act of 1964 and the Voting Rights Act of 1965, which transformed the landscape of racial equality in the United States.

The anti-apartheid movement in South Africa mobilised a range of individuals and groups, both within the country and internationally, to challenge the oppressive system of racial segregation. Ordinary individuals, such as student activists, grassroots organisers, and international supporters, contributed in various ways, from boycotts and protests to divestment campaigns and cultural resistance. Their cumulative efforts, alongside the leadership of figures like Nelson Mandela, contributed to the dismantling of apartheid and the establishment of a democratic South Africa.

The global environmental movement, driven by individual activists, grassroots organisations, and international collaborations, has had a significant impact on global environmental policy. Efforts such as the Fridays for Future and Plastic Free July movements, as well as campaigns against fossil fuel extraction, deforestation, and CFCs have raised awareness and prompted governments and institutions to take action. One notable outcome is the Paris Agreement, a global treaty aimed at addressing climate change, which resulted from collective efforts by individuals and organisations worldwide.

On a smaller scale, many of the everyday activists we interviewed for this book have made and are making transformative change. Thiha was part of the student uprisings in Myanmar in the 1980s that contributed to dismantling the authoritarian government and led to positive developments throughout the following decades until the sharp recent decline with the 2021 miliary coup. Phyo Phae Thida successfully lobbied the Australian Government to extend the visas of students from Myanmar based on the current precarious post-coup situation. Alicia's small startup in Papua New Guinea evolved into a private, public, and civil society partnership to support survivors of family and sexual violence and she has successfully influenced legislation and policy. Ben was a part of Australia's delegation to the United Nations that saw the ratification of an arms treaty in 2014. Alyssa is part of the movement to ban duck shooting in Victoria and had an important win in 2023 with a complete ban recommendation from the inquiry into native bird hunting, which received the largest ever number of submissions in Victorian parliamentary history (over 10,000). Kim led a successful movement to stop the subdivision of land that included a creek, wildlife corridor, and remnant bushland. Instead of 63 human dwellings, the land is now permanently protected and recognised as being of significant biodiverse value. Adrian and Janelle were heavily involved in lobbying and protesting against mining corporations and had wins related to this such as encouraging contractors working for these corporations to quit. Darrell founded an organisation that has now been in existence for 30 years. He contributed to the formation of a national park and continues to advocate on behalf of a diverse group of affiliates on local grassroots issues that threaten environmental sustainability. Delphine and her family supported Eddie Mabo's hugely impactful case against the State of Queensland in the early 1990s that led to the recognition of native title in Australia.

The cameos at the beginning of the book provide more detail about the numerous changes that everyday activists contributing to this book have achieved. The examples demonstrate how the actions of individuals, even at a micro-level, can have a transformative impact on global issues. They underline the power of collective action, the ability to inspire one another, and the potential to create social, political, and cultural change. While these individuals may have started with small-scale efforts, their actions resonated and mobilised others, leading to larger-scale movements and lasting global transformations.

The power of everyday people

Everyday activism refers to the notion that individual actions, no matter how small, can contribute to larger social change. It emphasises the importance of daily practices, behaviours, and decisions aligned with one's values and goals. Everyday activism recognises that even seemingly mundane actions, such as reducing one's personal carbon footprint, practicing ethical consumption, or engaging in conversations about social issues, can have ripple effects and inspire others to act. This does not negate corporate or government responsibility; it simply highlights that everyday people can positively influence the world in which we live.

Sociologists have unpacked the agency of everyday people and their ability to create change. For example, Ware et al. (2023a, 2023b) use the framework of everyday peace to capture progress made toward peace between Rakhine Buddhist and Rohingya Muslim villages in Myanmar where residents have made small individual efforts that contribute to greater cooperation, collaboration, and warm interactions between deeply divided communities. Everyday peace refers to the practices, interactions, and relationships that contribute to the maintenance and promotion of peaceful conditions at the micro-level of everyday life. Johan Galtung (1969), a prominent peace scholar, argued that peace is not merely the absence of violence but also the presence of justice, harmony, and wellbeing that addresses root causes of conflict, transforms oppressive structures, and promotes social justice. While positive peace is the goal, everyday peace theorists highlight the agency of individuals in creating and sustaining peaceful relationships and environments that cultivate a foundation for addressing systemic issues. Everyday peace theory recognises that individuals, through their everyday actions, can challenge violence, discrimination, and injustice, and contribute to building a more peaceful and just society.

On the other hand, everyday resistance theory explores the ways in which individuals and communities engage in acts of resistance and subversion within their daily lives. Political scientist and anthropologist James C. Scott (1987) coined the term *everyday resistance* to describe the informal, subtle, and often disguised forms of resistance that occur in various social contexts. Everyday resistance challenges dominant power structures and ideologies, even in the absence of overt protest or organised movements. Everyday resistance can take the form of acts such as non-compliance, evasion, sabotage, humour, or small-scale disruptions. These actions may seem insignificant on their own but collectively contribute to social change by contesting oppressive systems, asserting alternative values, and creating spaces for dialogue and transformation.

Structure and agency

The relationship between structure and agency is a central theme in sociology. Structure and agency theory examines how individuals operate within social structures. Structures set the context for action, while agency is about individual power. Sociologist Anthony Giddens' (1993) structuration theory illustrates how individual actions can lead to broader societal consequences. Giddens posits that people actively influence, not just passively exist within, societal structures. The interplay between structures and individuals highlights how people can be influenced by and have an influence on global issues through their choices and actions. Thus, everyday activists' actions can challenge established norms and introduce innovative avenues for change.

Collectiveness and cooperation

The advent of modern communication technologies has facilitated the diffusion of power and the ability of individuals to connect and collaborate across geographical boundaries. Social media platforms, for example, enable everyday activists to share information, organise campaigns, and mobilise support on a global scale. This diffusion of power empowers individuals to challenge existing structures and expand their impact.

Theories around collectivity and cooperation focus on our ability to link with others to achieve common goals (Deutsch, 2011; Rogers et al., 2021). For example, network theory highlights the power of connections, relationships, and collaborations among individuals and organisations (Borgatti et al., 2009). Engaging in networks allows individuals to harness collective knowledge, resources, and outreach, furthering their activism. Actor-network theory underscores the role of both human and non-human actors in determining social outcomes (Latour, 2005). This suggests that change emerges from interactions between various actors, including technology and institutions.

Similarly, social movement theory emphasises working together around shared objectives (Tilly, 2004). It is about empowering individuals to collaborate and drive societal transformations. Cultural transformation theories also highlight the importance of culture in directing individual actions and broader societal shifts (Fisher et al., 2011). Individuals can go against prevailing cultural norms by developing alternative ways of being and interacting; for example, using arts-based methodologies to influence collective perspectives.

In practice

The theories outlined above are not mutually exclusive and can inform and complement one another in understanding and supporting individuals to act for global impact. By drawing from these and other theories, we can gain a deeper understanding of the contexts, structures, and mechanisms that shape global problems, and the strategies and actions we can take to address them. Importantly, this includes critically reflecting on our roles within structures and actively working to dismantle oppressive systems in which we are complicit. By understanding the interplay between structure and agency, engaging in everyday resistance, forming networks, utilising modern communication technologies, and challenging cultural norms, activists can debate existing structures, shape social systems, and contribute to positive social change.

However, although the theories and discussion above can explain how individuals can act to influence large-scale change, they do not detail the practicalities of overcoming inertia in a problematic world. Noting that our fellow everyday activists often feel demotivated and disillusioned, we asked them how they overcome these feelings and continue with their activism. Many of the interviewees said they took time to identify their sphere of influence and focused only on what they can control. By understanding where they held power, they could narrow and apply their efforts in a targeted way. Feeling in control of personal actions, doing something small but valuable, and finding what they could do to contribute, helped the interviewees prevail against overwhelming feelings. Keeping the focus on the day-to-day activities brought the interviewees hope and renewed their strength to continue. The interviewees did not suggest relinquishing all thoughts of the wider issues, but rather advocating for others to address large issues that were beyond their control.

Another way of coping was for interviewees to reorient their thinking patterns. Some chose to focus on the future and the goal. Although this contrasts with the advice about keeping focused on the day-to-day, interviewees said that this helped bring perspective. Being future focused, acknowledging the long-term nature, and keeping a big picture perspective, enabled them to place their efforts into a global and historical context. Taking an expansive approach helped to remind the interviewees of their insignificance. It liberated them to think about what side of history they would like to be on and motivated them to be part of the solution rather than the problem. They noted that the personal cost of not doing something was too great; thus, they were compelled to contribute.

Interviewees also suggested reorienting thought patterns when it came to reminding themselves and others about the wins. They stayed optimistic in the face of demotivating factors by holding onto positive instances of change, no matter how small. They treasured moments of value along the journey and sometimes found ways of recording their results and using those to inspire others. They often let go of what they could not control but held onto the positives. Some interviewees suggested it was okay to assume that all wins were partially due to their personal contribution. Although retaining a sense of humility, they found value in claiming wins and letting the positive momentum carry them forward. These interviewees took any and every opportunity to hold onto and share positive feedback.

Some interviewees turned their negativity around a demotivating issue into a motivating factor. They chose to use their anger and frustration in a constructive way to overcome barriers and push forward. They gave themselves permission to feel overwhelmed. This then turned to anger. But instead of heading towards disillusionment, they channelled that energy towards further efforts.

Similarly, some interviewees applied a growth mindset when it came to making mistakes. Instead of giving up if something did not work, they saw mistakes as an opportunity for growth. They used these instances as learning opportunities to reflect on what could be improved and did things differently next time. Some undertook professional development, others read widely, and some journalled their issues to help find new solutions. If their actions did not come to fruition, they chose to pick up a different area of interest and try something new.

Many interviewees mentioned the importance of understanding the personal triggers that can cause inertia. Not having enough rest and working too hard were cited as common catalysts for demotivation and intolerance of people and issues that would not normally pose a problem. Interviewees emphasised that demotivation could be a sign of burnout. They also suggested avoiding things that lead to unhappiness or reducing exposure to the causes of disillusionment. One interviewee suggested containing the periods of negativity and not letting them consume their thoughts. Taking time out, travelling somewhere new, getting a break from the routine, and returning to people and places that recharge the batteries were suggested. Interviewees also recommended meditating, practicing gratitude, carving out personal time and space, and focusing on things they found beautiful and joyful as antidotes to feeling overwhelmed, as detailed in Chapter Eight on self-care.

For interviewees whose activism work was connected to their faith, they drew upon this belief system to help them through overwhelming times. Some considered their activism work to be their calling, so they felt unable to give up in times of demotivation or disillusionment. They believed the struggle was part of the process and what made the work worthwhile. They took negative feelings of demotivation and dis-illusionment as parts of a greater whole; as long as they were working on something they believed was right, the feelings were part and parcel.

Many interviewees mentioned the importance of social support systems for giving and receiving appreciation and solidarity. Family support, calling upon the wisdom shared by their grandparents, and drawing upon the strength they found when they thought about the family's commitment, were strategies they applied when feeling overwhelmed. Some interviewees tapped into their network of friends, connected with others sharing similar experiences or found people who share similar values. Finding strength in others who are likeminded and have a shared vision was important. These support systems could form naturally or they could be instigated more formally. For example, some interviewees suggested buddy systems as useful for supporting colleagues on an ongoing basis. A key message was the importance of accepting compliments and thanks from others. Marilee emphasised that we should absorb these compliments and messages of appreciation; hear them, feel them, and sit with them rather than humbly and quickly dismissing them.

In summary, the following strategies could help everyday activists overcome inertia and paralysis associated with addressing global problems:

- Take time to identify and think about the inherent structures that prevent or hinder change.
- Identify your sphere of influence and focus on a manageable locus of control.
- Focus on the goal and decide what side of history you would like to be on, noting that often silence is complicity.
- Treasure instances of change and positive feedback no matter how small.
- Channel feelings of being overwhelmed, angry, and frustrated into momentum for action.
- Understand mistakes and barriers as learning opportunities for growth.
- Identify personal triggers for demotivation and avoid them or take a break.
- Draw upon internal belief and social systems for support.

Conclusion

On a fundamental level, feelings of being overwhelmed are rationally linked to the layers of oppression that have developed historically to keep power within the hands of a privileged few. By reflecting on our position within the system we can identify our power and actively contribute to changing the system. Without doing this self-reflective work, and seeing how we are also harmed by the oppression of other entities, any humanitarian or activist action we take is likely to be from a place of pity, self-aggrandisement, or saviourism.

Returning to Lilla Watson's (2022) quote from Chapter One, "helping" is top-down; recognising how systems of oppression oppress us all, and then working to deconstruct those entanglements, allows us to work together for our collective liberation. The Song "None of us are free", sung at different times by Solomon Burke and Ray Charles, and written by Cynthia Weil, Barry Mann, and Brenda Russell in 1993, captures this ethic:

> If one of us are chained, none of us are free... If you don't say it's wrong then that says it's right. ...Now it's time to start making changes, And it's time for us all to realize, That the truth is shining bright right before our eyes.

This chapter emphasised the importance of recognising our place in the grand scheme of global problems and the significance of taking action, despite our individual limitations and transient existence. By drawing insights from systems thinking and sociology, we have explored the intricate relationship between societal structures and personal agency, acknowledging the intersection between the systems we inhabit and our capacity to affect change.

We have delved into the notion that ignorance may provide temporary solace, but ultimately hinders our progress and prevents us from fully engaging with the world. Unpicking our place in the complex dynamics of society uncovers a web of privilege and disadvantage. Acknowledging our complicity in systems of oppression illuminates the imperative for action. Layla Saad (2020) while talking about white supremacy, made a point deeply relevant to all global problems when she said:

> *You* are part of the problem and...you are *simultaneously* also a part of the answer. There is great power and responsibility in that knowledge. But knowledge without action is meaningless. To dismantle this system of oppression and marginalization that has hurt so many for so many generations, we need all of us. In creating a new world, everyone's contribution matters. ...No matter who you are, you have the power to influence change on the world. The effects of your actions, whether consciously chosen or not, will

impact everyone who comes into contact with you and what you create in the world while you are alive.

(p. 210)

Rather than vilifying ordinary individuals as bad actors within structural systems, this chapter has introduced sociological theories that highlight the potential impact that individuals can have through networking and collective pooling of individual micro-actions. These everyday actions are collectively capable of driving change on local, national, and global scales. This chapter serves as a call to action, encouraging readers to embrace their agency and contribute to the ongoing transformation of societal systems. It is through the recognition of our interconnectedness, the understanding of our complicity, and the mobilisation of collective agency that we can challenge and dismantle oppressive systems. By doing so, we have the capacity to foster a more equitable, just, and inclusive world for all.

References

Borgatti, S. P., Mehra, A., Brass, D. J., & Labianca, G. (2009). Network analysis in the social sciences. *Science*, 323(5916), 892–895.

Deutsch, M. (2011). Cooperation and competition. In P. T. Coleman (Ed.), *Conflict, interdependence, and justice: The intellectual legacy of Morton Deutsch* (pp. 23–40). Springer. doi: https://doi.org/10.1007/978-1-4419-9994-8_2

Festinger, L. (1957). *A theory of cognitive dissonance*. Stanford University Press.

Fisher, R., Ury, W., & Patton, B. (2011). *Getting to yes: Negotiating agreement without giving in*. Penguin.

Galtung, J. (1969). Violence, peace, and peace research. *Journal of Peace Research*, 6(3), 167–191. doi: https://doi.org/10.1177/002234336900600301

Giddens, A. (1993). *New rules of sociological method: A positive critique of interpretive sociologies*. Stanford University Press.

Latour, B. (2005). *Reassembling the social: An introduction to actor-network-theory*. Oxford University Press.

Lilla: International Women's Network. (2022). *Lilla: International Women's Network*. https://lillanetwork.wordpress.com/about/

Mann, B., Weil, C., & Russell, B. (1993). *None of us are free*. https://en.wikipedia.org/wiki/None_of_Us_Are_Free

Rogers, A., Kelly, L., & McCoy, A. (2021). Using social psychology to constructively involve colleagues in internal evaluation. *American Journal of Evaluation*, 42(4), 541–558. doi: https://doi.org/10.1177/1098214020959465

Saad, L. (2020). *Me and white supremacy: How to recognise your privilege, combat racism and change the world*. Quercus.

Scott, J. C. (1987). *Weapons of the weak: Everyday forms of peasant resistance*. Yale University Press.

Tavris, C., & Aronson, E. (2007). *Mistakes were made (But not by me): Why we justify foolish beliefs, bad decisions, and hurtful acts.* Harcourt.

Tilly, C. (2004). *Social movements, 1768–2004.* Paradigm Publishers.

Ware, A., Ware, V., & Kelly, L. (2023). Everyday peace as a community development approach. In A. Kilmurray, J. Eversley, & S. Gormally (Eds.), *Peacebuilding, conflict and community development* (pp. 25–39). Policy Press. https://doi.org/10.1332/policypress/9781447359333.003.0011.

Ware, V., Ware, A., & Kelly, L. (2023). Everyday peace: After ethnic cleansing in Myanmar's Rohingya conflict. In A. Kilmurray, J. Eversley, & S. Gormally (Eds.), *Peacebuilding, conflict, and community development* (pp. 191–208). Policy Press. doi: https://doi.org/10.1332/policypress/9781447359333.003.0011

3 Exploring underlying motivations

This chapter examines what specifically compels some people to become activists. The previous chapter focused on how to overcome inertia in a world where the problems have deep historical roots and are entrenched within systems of power. We now unpack why, for the people interviewed for this book, ignoring the problematic world was just not an option. They were driven by a force that would not allow them to ignore injustices. While all were compelled by their beliefs and values, for some it was linked to their past or present experiences and the situations in which they found themselves. For others it was linked to empathy, passion, a sense of responsibility, and sometimes guilt. Organised religion provided inspiration to several interviewees, with representation from Hinduism, Buddhism, Islam, and Christianity, among others. Some interviewees were allured by the notion of legacy and the desire to be a good ancestor.

Interviewees felt an intrinsic pull to act, which was captured throughout the interviews. Alyssa felt that "there really isn't an alternative – it's the least we can do." Khalil explained that: "if I see an injustice, I want to do something about it." Others highlighted a personal calling: "I knew I was in this world…to do something worthwhile" (Alicia), and "this is really my calling and this is the point at which I am to serve" (Serena). Mike captured the importance of attempting "as far as possible, to not add to the pain and suffering in the world. To try and be aware of my own motivations; the volition behind each action… to attempt to be a kinder person." Whatever people's reasoning, for many activists, ignoring the problems of the world is not an option. Whether it is our experiences, knowledge, values, or sense of responsibility that drives us, we accept our role as actors on the Earth and seek to use our time to contribute towards making the world a better place.

There is inherent value in understanding personal motivations and being able to articulate these clearly. Taking the time to question,

DOI: 10.4324/9781003333982-3

consider, and articulate why you contribute to saving the planet is a valuable step in personal development. Thinking about what a good life entails, what gives your life meaning, and why you feel this way is an integral part of developing wisdom about yourself and what it means to be human. This chapter provides a prompt for this intro-spective journey. Although there are different reasons why individuals are motivated to contribute, working on something that they enjoy and is bigger than their own concerns can help with finding a sense of peace, meaning, and fulfilment. Interrogating motivations using psy-chological and philosophical theories of human motivation may help to determine whether actions are helping us live a meaningful life. The chapter concludes with a set of reflective questions that may help elu-cidate underlying motivations and understand whether those actions contribute to a sense of fulfilment.

Why are activists compelled to act?

The people interviewed for this book recounted multiple motivations for why they choose to contribute to activist initiatives. Many of these motivations overlap or wax and wane over the life course. Patterns and links between the various motivations identified the following key themes: personal experience, a desire for justice, duty, guilt, under-standing, empathy, values, faith, and legacy.

Experience

The main reason provided by the interviewees for why they were con-tributing was because of a childhood or personal life experience. Lived experience is an important motivator that can provide people with deep knowledge and compassion for others experiencing similar situations. Often these life changing events were when they had witnessed or experi-enced conflict, inequality, oppression, injustice, trauma, or discrimination. Lived experience of adverse situations helped some of our fellow activists develop an intense sense of justice and anger to propel their initiatives.

Kim said that her passion for justice was cultivated through lived experience of injustice and discrimination whereby she was regularly racially vilified growing up in England. She recognised the falseness of those vilifications that alleged she was a dirty thief or worse for her Romany/Gypsy surname. Knowing the accusations to be untrue helped her see the untruth facing others and compelled her to address it: "If I see an injustice I want to try and help fix it." Darrell highlighted that lived experience of adversity growing up in poverty during the depression resulted in him feeling that he had no recourse except to act: "I had no choice. I had no choice." Domuto also shared how his life experience of inter-generational trauma in South Sudan during the civil war shaped his work:

> The violence of war affected me even before I was born. My parents and grandparents suffered from war. I was born in conflict, grew up in conflict, and work with conflict every day. I experienced the death and suffering of my parents. Wars and conflict prevented me from going to school many times. It blocked many opportunities. It robbed me of my childhood. It divided my family. There was a time I could only think of revenge. Taking a rifle and learning how to shoot. My mother helped me, education helped me, and changed my journey towards peace. My experience of conflict has shaped my life and commitment to bring people together. My journey has not been easy, but I have learned that other people do not have to suffer because I suffered. If we can unite the youth, we can build our country together. Then no one has to suffer.

For Ekawati, others' lack of belief in her ability to succeed drove her to push for change. "Growing up, I mostly tried to prove myself and prove my parents, neighbours, and relatives wrong. They used to tell me that I won't amount to anything since I can't hear. That's how I developed a rebellious and loud character." As a result of the low expectations placed on her from childhood, Ekawati pursued personal success but also strove to inspire other deaf and hearing-impaired children to reach for their dreams.

Ricki spoke about the abuse she received from her foster family and others in her community for her Indigeneity and queerness and how that crystallised into resolve. She grew up being taught "that any deviations from the binary model of male and female were wrong and needed to be fixed." She "was referred to as 'that thing' – 'I don't want to sit next to that thing, be near that thing, I don't want to be associated with *that thing*.'" Despite a long and difficult road of discrimination and exclusion, she explained how "All of these things formed me and I started breaking out in little ways." She highlights that her story is "my lived experience and how I've weaved it into making something happen… they laughed at me. But now I sit proud." Ricki spoke similarly of her disabilities noting that:

> It's about building me and showing that I have a voice, and showing the world that you can do it, no one can stop you. It doesn't

matter what disabilities you have; it only becomes a barrier if you let it become a barrier. If you know how to use the system you can advocate for yourself, find likeminded people, you will get places.

For some interviewees, part of their healing journey was helping others avoid suffering and supporting them to recover. Alicia said:

I come from a poor background… All of that drives me, but the work in the violence space was because I was a victim and I felt it. And freedom, my freedom from that relationship is the reason why I do everything today because I'm free to help other people.

Noraini said, "I'm more and more believing the point that when you're helping people, you're helping yourself. When you're helping people heal, in essence you're healing yourself."

These beliefs resonate with the work of Judy Atkinson (2002), author of *Trauma trails – Recreating song lines.* She highlights the value of refocusing energy that arises from personal experiences: "The only power I have is to journey with those in the present, to take the opportunity we have now to make our country a safe and healing place to live, for our children and grandchildren" (p. 266). This aligns with investigations into post-traumatic growth which have found that activism and meaningful engagement can assist with overcoming interpersonal trauma and contribute to breaking cycles of intergenerational trauma. Holocaust survivor, renowned physician, and addiction expert Dr Gabor Maté writes, "Our concept of well-being must move from the individual to the global in every sense of that word" (Maté & Maté, 2022, p. iii).

Motivation from personal experiences includes positive events and encounters. Mehreen was raised by her grandparents and noted that this resulted in her "affinity for older people". Talia explained that she "started a business when I was eight making hair ties and hair clips and jewellery to make money for Amaze for autism because my little cousin was diagnosed with autism." Talia grew up in a community where environmental action was important. She said, "It opened my mind to problems that were bigger… I found it really interesting and spent my spare time researching things and became passionate about it and wanted to make my contribution – knowing I'm doing my bit."

Despite common links to personal experience, many people activate around issues for which they have no immediate connection. However, links with lived experience influence our choices in ways that are not always clear. For example, an experience of family violence may draw an activist to develop an interest in supporting women and children.

Further, our lived experience shapes who we are, the depth of our empathy, and our ability to understand complexity. To summarise, Roy provided an explanation for how this theme of personal experiences can generate change on a larger scale: "I think that we all, as a result of our life experiences, we will take on a passion for something, me something, you something else, and together we will get it all done."

Justice

Former Australian soccer captain Craig Foster is a human rights advocate for asylum seekers. In *The Big Issue*, he shared how he has always had an interest in social justice:

> A drive for justice and equity was always present in my life... from a very early age. I had a visceral dislike of arrogance or the feeling that someone considered themselves superior to others and entitled to different treatment... Humility and seeing others as equal were always very important to me.
>
> (Foster, 2022, p. 19)

Some interviewees also discussed the importance of upholding fairness, justice, equity, and moral principles. Helping others (human or non-human) who were experiencing an unfair predicament that stopped them from living a full existence was an intense motivation. Linked to strong social justice values, many activists feel compelled to act in solidarity and support for others as they feel it is their responsibility as a good citizen. They could not sit by and let things pass.

Some interviewees similarly noted that they were aware of injustice from a young age and always felt a responsibility to act. Noraini said, "If something didn't feel right, if it didn't make sense, I had to do something about it." Kim said she had, "an innate sense of justice for people and recognising an absolute abhorrence for discrimination, for arbitrary discrimination... That's the key, we feel responsible. I definitely feel that." Khalil also felt a need to contribute from a young age:

> There are so many injustices in the world and when I was younger I saw them but I didn't know what to do. I learned all my life to try and improve myself, to be a better person. I am learning how to fight [injustice] more. And this is the whole meaning of life. How I become better and better at working to change reality and address injustices... Speaking up is very important for me, to give meaning to life.

Ben said when he was younger he saw an injustice on television related to the misrepresentation and mistreatment of asylum seekers and from then on:

> I needed to do more. There was real horrible injustice happening in the world that needed opposition to it... So, part of my motivation was to learn as much as I could about the way the world works and history and the kind of frameworks that progressive people use to understand the world and envision a better one... It just feels like you can't really do anything else. It's not really a choice to think differently, or to act differently, or to put your head in the sand about this stuff because it's just not viable.

Connected to acknowledging universal human rights, some interviewees said that if their actions do not harm others, then they had a desire to ensure that others have choices and the freedom to live their lives the way they want. Nafiz feels a sense of duty towards his community in Bangladesh and that fuels his commitment to work towards positive change. Iram said:

> I believe in equal opportunities for all. My ideal society is where nobody is rejected, feels insecure or is treated unequally. Every living being has a right to exist in a most respectful and dignified way. For me, denial of rights to one individual is denial to all humankind. I see the capabilities of individuals, not who they are and where they come from.

The complacency around injustice bothers Alicia, who comes from a poor background where education was her pathway to freedom. She empathises with those who lack opportunities and require assistance. Alicia sees herself as someone who notices and supports those who are overlooked by others. Her determination to make a difference leads her to ask questions, take action, and persistently advocate even when her ideas are shut down. Similarly for Nicole, sitting on the sidelines and watching injustices unfold is not an option. She feels an obligation to be part of the solution, actively engaging and contributing to create a better world. Nigel feels that he cannot say no when there are people relying on him to keep going. This expectation drives him to continue his efforts, refusing to turn a blind eye to the issues at hand.

This motivation aligns with the thinking of the French philosopher, mystic, and political activist Simone Weil who believed that justice was about obligation (cited in Rozelle-Stone & Davis, 2023). She was a moral philosopher who wrote that justice was about making sure no harm is done to another. Weil understood that humans are fundamentally unconditionally obligated to uphold justice based upon a duty to the very nature of humanity. Essentially, her argument is based on the age-old premises of the importance of treating others as you would like to be treated:

> Only the absolute identification of justice and love makes the coexistence possible of compassion and gratitude on the one hand, and on the other, of respect for the dignity of affliction in the afflicted—a respect felt by the sufferer himself and the others.
>
> (Rozelle-Stone & Davis, 2023, p. 1)

Duty

In a featured article in *The UNESCO Courier*, Nigerian American academic Michael Onyebuchi Eze (2011) writes about *Ubuntu/botho*, which is translated from the Bantu language into *I am because we are*. Eze (2011) writes:

> Ask anyone on the streets of Harare, Johannesburg, Lusaka or Lilongwe (in Southern and Eastern Africa) what they understand by Ubuntu/botho and they will probably list the virtues to which a person in these societies is expected to aspire – such as compassion, generosity, honesty, magnanimity, empathy, understanding, forgiveness and the ability to share.
>
> (p. 1)

In this theme we cover how some interviewees felt they had a duty to aspire to these virtues. They felt morally bound or obligated to contribute. Activism for these individuals is an act of respect.

Interviewees suggested that power, education, access, and skills come with a responsibility to generate change towards a fair society. They demonstrated a deep sense of responsibility and a duty of care towards the Earth and its inhabitants, even when advocacy and activism may put them at risk. They spoke of others relying on them. They needed to be "part of the solution" and not just "sit on the sidelines" (Nicole). As well as recognising their responsibility to others, they recognise their responsibility to themselves. The way we use the level of privilege and opportunity afforded to us impacts our self-respect and identity.

Elise commented that she wants to, "use my privilege and my power in whatever small way possible to help address those problems [of inequality]". Nigel said, "I've always just felt very strongly that I've had a very privileged upbringing and free higher education and I just think that you shouldn't waste that. You've got to put something back in." Adrian also noted, "I've got time and talent on my hands." When Serena was at university she reflected upon her background, recognised her comparative privilege, and had a revelation, "if I can combine my education, my knowledge and my skills, even the privilege I came with, this idea of creating social change, that's what I'm going to do with my life". Other interviewees said they were motivated to use their privilege, apply their nuanced understanding of the underlying causes of injustice, and utilise their advantages to assist towards pro-social goals.

In South Sudan, Domuto feels a sense of responsibility that if he does not risk his life in pursuit of peace, his people will continue to suffer in ways that will lead to increased poverty, loss of property, and more individuals becoming widows and orphans. This understanding pushes him to overcome fear and be brave. For many people working in dangerous situations like this, their sense of duty to act outweighs their fear of repercussion. This is also the case for Phyo Phae Thida who sees it as her duty to raise awareness about the challenges her country and its people face. She actively connects with universities and civil society organisations, using her role as an emerging academic to inform the international audience about the issues occurring in Myanmar. Despite compelling safety reasons for silence, she regrets having stayed silent in the past, considering it a betrayal of her values, and is committed to never repeating that mistake.

As a privileged son of an army colonel, Thiha could have chosen an easy path, but his sense of social justice compelled him to act, despite dire consequences that saw him incarcerated for decades then forced to flee his homeland. While his activism deeply affected his life and has resulted in significant trauma, Thiha explained that he "had to do something" because "it was very unfair" and he would have found it hard to live with himself if he had closed his eyes and stayed in his "perfect" life. Thus, taking action, speaking up, and advocating for change to create a more just and equitable world represents enaction of a duty to himself and to others in his life. Fortunately, most everyday activists do not face the prospect of death for their activist actions, but this extreme example of confronting risk demonstrates the intense pull of our convictions to act. Phyo Phae Thida suggested that "everybody has the responsibility to do something" and that cultivating this sense of responsibility in others can help bring them on the journey.

Guilt

Some interviewees indicated that part of their motivation was linked to feelings of guilt. They felt guilt in relation to taking up space, utilising finite resources, and feeling a sense of privilege. Despite Alyssa's extensive activism in support of native wildlife, she commented that it is "hard to think of what I do to make the world a better place. I do a lot of things that make the world a worse place." Ben was concerned that "there's definitely a feeling always that I'm not doing enough or I'm not doing the right things. To an extent, I guess, sometimes I try and do more, and sometimes... it sits in the back of my mind subconsciously – a nascent feeling of guilt that never goes away." Elise mentioned: "I sometimes feel that I'm not doing enough or that I've become kind of a bit lazy in my activism."

The activist actions of the interviewees helped mitigate some of their negative feelings of guilt, pain, fear, rage, and shame. Noraini explained, "At the end of the day... you know that you've made a bit of a difference, even just a dent in your own small surroundings. I'm not just here taking up space." Matt shared, "It feels good to at least be trying to make a positive impact. I feel like, given the state of things I'm obligated to try and do whatever I can personally."

Avoiding guilt was also mentioned in relation to just existing on the planet. Being alive, using resources, and contributing to ongoing harm makes some interviewees feel guilt and prompts them to contribute. For Tamsin, she finds that contributing stops her from feeling "like a big blob that's just using up all the resources. I've earnt my place here." Along a similar sentiment Alyssa said:

> I honestly think a lot of it comes down to my general guilt for being alive. I have a genuine sense of displacement and that I shouldn't be here. So, I'm trying to make amends. That's a big motivator, a general feeling of guilt and hopelessness, and wanting to see some kind of future.

Documentaries that make Alex feel guilty are also a catalyst that prompt her into action:

> Any time I watch any documentary there's still a little bit of guilt will creep back in... the guilt comes from – we really have stuffed up the world – I'm a person in the world so some of the onus is on me. By doing steps and trying to engage others and going beyond my own environment to make an impact in spaces that I walk in, it certainly has felt that I'm making an impact and that I can sort of rest a bit easier – take the burden off myself.

Contributing can assuage guilt. Ekawati said, "Do I engage in daily activism because it's the right thing to do or do I want to sleep well at night?" This is reiterated by Nafiz: "When I go to sleep, I can have a peace of mind... I nearly always choose to work for betterment of people." Similarly, Liz said, "I think for me it makes me feel good... And it gives me a lot of a sense of contentment, just being able to see how I can contribute to changing somebody's life positively... When I sleep, I feel very content that I have been able to change a life."

Research has shown that feelings of guilt in the environmental activism area have led to positive behaviour changes (Antonetti & Maklan, 2014; Culiberg et al., 2023; Lindenmeier et al., 2017). Guilt may also have a role to play in many of the current topics facing our society, including recognising the victims of historical atrocities, colonialism, the environment, climate change, racism, and feminism (Buschmeier & von Kellenbach, 2022). However, feelings of inadequacy are common. Some activists can hold themselves to such high standards that they are unable to meet them, which can result in burnout (Ma'anit, 2007; Peace News, 2009; Utt, 2014). Chapter Eight on self-care speaks more about burnout, but here we argue that guilt is largely an unhelpful emotion that should only be harnessed as a motivational force to right wrongs if exercised in combination with self-compassion (Ágoston et al., 2022). For guilt to become a useful mechanism for lasting change we need to understand how to harness guilt appropriately to positively repair damaged relationships, balance power, and regulate social relations (Buschmeier & von Kellenbach, 2022).

Knowledge

Becoming aware of an issue can have a fundamental impact on the learner and result in individuals taking action. Knowledge challenges ignorance and opens our minds to the pressing issues that exist. Education and awareness, acquired through formal education, personal experiences, or immersion in an unfair or challenging context, are instrumental in compelling individuals to act for a cause. Although people can become desensitised to information about the extent of global issues, for some of the interviewees ignoring the problematic world is not an option. They spoke about how knowledge and awareness prompted them to confront issues and work towards solutions. As noted by Alyssa: "Once you've seen it, you can't unsee it."

Marilee explained how education shattered her preconceived notions. She was born and raised in a conservative Christian family, unaware of the realities beyond her small heteronormative town. However, her encounter with a book about modern-day slavery ignited a desire for social justice, leading her to study human rights and law.

This newfound knowledge compelled Marilee to reject many of her previous beliefs and actively fight for progressive causes. Her transformation showcases the power of education to inspire individuals to engage with the world's challenges.

As well as formal education, if our ears and eyes are open, our informal education will be immensely rich. Awareness can be raised through firsthand encounters, which can broaden perspectives and make individuals more empathetic. Elise and Roy's travels overseas opened their eyes to global challenges, such as poverty, inequality, pollution, and marginalised groups' struggles. For Roy, this exposure prompted him to reflect on his privileged middle-class life and limited worldview. He recalled, "I came back to Canada around Christmas and couldn't stop thinking about these kids and it really rocked my world." For both Roy and Elise, these experiences ignited a passion and sense of responsibility to act.

Extending this notion, Phyo Phae Thida highlighted that those with the privilege of access to knowledge should not only act on that information but also share it with those who do not have access. Speaking of people in her native Myanmar, she explained:

> I had the privilege to learn and assess critically and see the situation from a different angle, but so many people back home don't have that opportunity to learn and get information from different places. So I see that I have an obligation to share back to my community so that they can get that information as well…it's important for social change. If we really want to change our community, you can't do it alone, you have to join with other people. But in order to get others to join you, you have to raise their awareness. The first time they might not listen but if you keep talking and they keep hearing the same message again then maybe somehow they will think that they want to join you. Even if they don't change, at least they're aware.

Many interviewees recounted how social media was not only a powerful tool for sharing information with others but in some cases something they had read on social media had ignited passion for a cause. Although social media can be a resource for knowledge, of course it is also a vehicle for spreading misinformation (Marin, 2021; Motta, 2018). Regardless of the source, thinking critically about new information and being mindful of what sources are influencing our beliefs is critical. We all have more to learn, more to re-learn, and more to unlearn. For example, scholarly work on decolonisation provides useful information to help us understand our inherent biases and unpack our preconceived beliefs (e.g., Land, 2015; Smith, 2012). Hence, new information can be enlightening and a prompt for action, but it can also make things more complicated and uncertain. The journey is ongoing as shared by New Zealand Pākehā (non-Indigenous) evaluator Kate McKegg (2019) who is working towards unpacking white privilege needed to support Indigenous sovereignty and self-determination, "It can be exhilarating and also deeply challenging because our dominant knowledge systems and structures are powerful and resistant to allowing the necessary time and space for equitable co-existence of other knowledges" (p. 362).

Empathy

Buddhist writer Joan Halifax (2018) understands empathy as something that can help us sense what is going on with another person, make us feel close, inspire us to act, and help us understand the world in greater depth. Scientists have described how empathy has inspired their work to advocate for the environment, such as Stephen J. Gould who wrote in 1991: "We cannot win this battle to save species and environments without forging an emotional bond between ourselves and nature as well – for we will not fight to save what we do not love" (cited in Boon, 2021, p. 1). A leading neuroscientist, Professor Anil Seth (2021) researches how empathy provides humans with "an ability to see themselves as less *apart from*, and more *a part of*, the rest of nature" (p. 16).

Interviewees noted a deep respect for and connection to nature. This motivated them to want to take care of their environment and the beings that inhabit those lands. Kim described that she feels "A real sense of wonder and oneness with nature, that inspires me a lot... I mean how can I explain that feeling I have when I recognise and respect the intrinsic value of everything in nature?" Tanya said, "I have such a strong connection with animals and I have such a strong feeling of empathy whenever I see an animal struggling or suffering."

For Aruna, witnessing the plight of others and feeling empathy was directly linked to her actions:

> It was a decade long conflict and I could see young people being pulled out from schools and brainwashed. Which really made me think what is going to happen to these young people when the situation calms down? Are they going to be visible? Are they going to be listened to? Are they going to be heard? What's going to happen to them in the community?

Dia describes empathy as a motivational factor because she understands it to be something that flows between beings: "The empathy that I give and receive. I've created that sort of exchange. That karmic circle that comes back to me. It makes my life meaningful." Iram indicated that it is important "To be kind to nature, birds, pets, and of course human beings around us. I believe that in the coming 20–30 years, empathy and kindness is what will drive the world."

Some people intrinsically feel empathy from a young age. Ricki became tearful when she passionately said, "I just care about everything so much and everything means so much for me." Putima said, "I didn't like seeing people from a very early age being upset or scared. And I loved sharing… I think I was just born very sensitive." However, there is psychological evidence emerging that empathy can be taught. It seems that there is not a clear delineation between people who are empathetic and those who are not. Jamil Zaki (2020) is a professor of psychology at Stanford University and the director of the Stanford Social Neuroscience Lab. His book *The war for kindness: Building empathy in a fractured world*, examines how empathy works and shows that it is a skill that can be developed. We cover this in more depth in Chapter Five on cooperation.

Authenticity

Some of the interviewees discussed their desire to live an authentic life and shared notions of reliability and genuineness. They spoke of how once they had become aware of the issues at hand, it became crucial to connect the information with thoughts and feelings and translate them into action. Merely knowing something was not enough. They did not want to distance themselves, resort to cognitive dissonance, or make excuses to justify inaction. Interviewees felt that it was important to stay true to themselves, maintain their values, and avoid living hypocritical lives to the best of their ability.

For many of the interviewees, the idea of living a blinkered and ignorant life is unappealing. Mike mused on this way of living as inauthentic. He observed that those who accepted ignorance do not seem to lead fulfilling lives. Mike does not want to live a life where he trades comfort for awareness and engagement with important issues. This drives him to maintain reflexivity and keep pushing forward. By acknowledging the impact of our choices, staying true to our values, and challenging internal conflicts, we may live fulfilling lives and make a positive difference in the world.

Recognising that our choices as individual consumers have a direct effect on larger issues, such as environmental concerns or ethical practices, can be a powerful motivator. Elise highlighted her strategy of focusing on issues within her sphere of control, acknowledging the power that lies within individual choices. She explained: "It feels kind of narcissistic in a way because it's like you're just focusing on issues that affect you, but in other ways, that is where you as an individual hold a lot of power." By being conscious of our consumer decisions, we can contribute to positive change and shape a better future.

Values are an integral part of our identity and self-worth. Many activists strive to align their actions with their beliefs to avoid being hypocrites. Veganism provides a case example to explain the concept of authenticity. It entails abstinence from consuming or using any animal products based on the ethical position of extending social justice to non-human animals (Singer, 2009). Many vegans understand veganism to be a social movement and an identity. As exemplified by eight of the 46 everyday activists interviewed for this book, their commitment to veganism reflects their values. They recognise that fighting for rights while causing suffering for the sake of personal pleasure would contradict their principles. Confronting and excavating their deeply held beliefs was a crucial step in the process of living a life with purpose. Tanya's decision to quit horse riding after becoming vegan reflects the internal conflict she experienced. The inconsistency between her values and her previous actions led her to critically consider the animal's rights. "It didn't feel right anymore" she explained. Similarly, Jenny's journey towards veganism was initially met with (self and external) resistance due to her upbringing on a cattle farm and social pressures. However, the discomfort of not being true to herself and her beliefs, ultimately motivated her to break free from dominant narratives and align her actions with her values.

The journey of life presents challenges and obstacles, but it is important to be kind to ourselves while remaining steadfast in our convictions. Adrian captured this sentiment: "I'm really, really concerned to walk my own talk as much as I can, knowing that I'm human and I'm going to fall occasionally and slip and trip, despite my best efforts at maintaining my balance." Similarly, we change over time as our knowledge increases. We might look back on our actions last year or last decade and shake our heads. That is okay though. If we are doing the best that we can at the time, striving to grow and learn and improve, that is success.

Faith and spirituality

This theme relates to faith as system of religious belief and connects to spirituality as something that is incorporeal or immaterial nature. While we did not specifically ask about religious affiliations or spirituality, over a third of interviewees raised their alignment to religions including Hinduism, Islam, Catholicism, Christianity, and Buddhism. They connected their religion with addressing moral aims such as being good, performing good acts, and trying to make progress on social and political goals. Often their profession or work as an activist aligned with their religion.

Several interviewees mentioned terms such as integrity, passion, and generosity. Doing good, upholding family and community commitments, giving to others of their time and other resources, and acting in a way that you would like others to act towards you, were strong sentiments. Delphine said, "I'll just ask God to give me strength to do this... Even if I don't feel like doing it, it's the right thing to do and therefore that's what love is. Love is doing the right thing by anything and anyone that is living." Similarly, Liz said, "My parents believed a lot in the Christian faith... issues around integrity were really big for my mum and the values around helping other people... reach out to other people and support them... I'm always looking out for ways that I can support other people." Other interviewees described having a calling that connected their religious or cultural practices. Serena said:

> My Christian faith is a very big part... From a system of faith that also comes back again to culture... my place is here so that I can serve – that's the Bible that I studied and the faith that I have been introduced to. And it's always been very much embedded and ingrained in me that this is really my calling and this is the point at which I am to serve.

Many interviewees were also intent on upholding ideas, values, and maxims linked to the religious teachings they learnt as children. Mehreen, being raised and embedded in the Canadian-Pakistani Muslim community, said:

> There's a word in my culture, *sewa*,[1] which means to give... I would definitely say I was raised in a home about volunteering and my faith is based on service and giving back. You should do what you can do for others. So, it runs really deep in me for sure. Selfless service coming from a place of giving to give.

Even though Susanne is no longer a practicing Christian, Christianity and Buddhism influence her value system. Susanne said:

> I still think one of the underlying things that I liked about Christianity is working with marginalised and vulnerable people. I have a

lot of privilege and I want to use that to walk alongside other people... Love guides the way I work – love as its connected to justice and fairness, and human rights. I wouldn't want to be doing anything unless it aligned with my values and was creating change. I wouldn't say Jesus is my motivation, but if he was a real dude, I do like how He operated in the world. I'm not Buddhist, but I am motivated by those thoughts of gratitude, compassion, and loving kindness.

Similarly, Marion said, "My values and beliefs are around Buddhist principles about being responsible for my own life, being responsible for helping others, and trying to make the world a better place. So I'm keen for there to be a society in which we live that has values like equality, fairness, and justice in society for all."

In an article in *The Monthly*, Don Watson (2022) discusses how King Charles III is profoundly religious and seeks a universal commonality, not only within the Church of England and Scotland, but also eastern Orthodox mysticism, Buddhism, Islam, and Greek Orthodoxy. King Charles has a strong record as an environmentalist and Watson (2022) makes the connection between his beliefs within Traditionalism and his understanding that humans are participants in nature, as opposed to conquerors of the natural world. Paganism also has strong themes connecting with and protecting the environment. Jóhanna Harðardóttir, a priestess of Ásatrú, an ancient Icelandic pagan religion, says that Paganism orients her in nature: "We are, of course, nature. We are born from nature and we will go back there – and while we are here in between, everything we have and everything we will ever own comes from nature" (Harðardóttir, 2021, p. 1). Aligning with a common sentiment among our interviewees, literature is continuously emerging that connects environmental issues to spirituality and helps us understand how many humans see themselves in relation to where they live (Allison, 2015).

Legacy

Interviewees spoke of their respect for ancestors and their desire to leave a legacy of having acted in accordance with their values for future generations. In relation to respect for generations past, many interviewees had at least one family member who was an activist or inspired them to contribute in some way. Mosrat said, "Two of my family members are actually working in this sector and from their storytelling they actually involved me [in discussions around] how the people are surviving... and how there are people contributing to improving the community's life." Some interviewees described this as a form of cultural obligation. Ronny was inspired by his mother's explanations of connection and obligation to others, which was immortalised in her motto: "If I'm hungry, you all go hungry. If I eat, we all eat." Speaking of West Papua, Ronny said:

> Our Melanesian way of the cultural connection with the land, the language, the songs, the kinship system, and the respect to the elders, some of the norms around how you engage as well in conflict resolution… our cultural practices… connect us… The values and cultural practice that come with it [contributing]… remain very strong.

Strong community and familial links perpetuate social bonding that is boosted by cultural and generational expectations to act for good. Serena saw herself situated between generations past and present with an obligation to continue the work of Elders to inspire the young. Aboriginal and Torres Strait Islander women Delphine and Putima shared that they come from activist families and supporting their family and community is a duty they wanted to uphold. Delphine said:

> I probably could say I was born into it. Both my parents were activists without even trying. My mum because of her story and how she personally responded to her situation. I think in hindsight, it's basically to honour them because my mum's life was really hard.

Delphine highlighted that this is about focusing on those in need and not indulging in self-pity. She said, "You can't look down here and look in the mirror and see yourself and say poor me. You can't do that because this is a crucial and critical time – When others need you, you help." Putima said that she comes from generations of activists and can trace her lineage through the actions over decades, "I come from a very big family. We're a family of activists. I was very lucky that I wasn't sheltered from the stories. Because that makes up our family and tells the history of the struggle." Putima's grandparents continued their activism into old age. Their legacy and example meant that she felt she could not give up. She summed up her vision in this way:

> I believe in human rights and Indigenous rights and workers' rights… People having good lives – being able to have choices and get help when they need it. And the basic thing is I like people to be happy, I want

people to have a happy life and feel if they want to contribute they can, if they just want to be themselves, just being, they can do that too… Equality, I want people to have the choices and have the same opportunities as everybody else. I picture my world and I can picture that everybody has a chance to get supported to do what they want to do.

For some interviewees, they shared what they wanted to accomplish whilst they were alive because they were acutely aware of their mortality. They did not necessarily fear dying, but acknowledged how short and insignificant their time on this Earth is in relation to the universe. Jenny shared how death motivates her to live a full life:

I've had quite a bit of a fear of death since I was quite young… We've got one life, but every day we have the potential to leave something good in the world… Love, connection, stories, and narratives, of compassion and kindness, it's our legacy. Every person matters so much. When we start the dialogue about death, then we can live more richly.

Vijeta summarised her thoughts in this way: "Your time is limited on this planet, so I want to live a meaningful life. Each day is a blessing and I need to make use of that… Our time is limited so making full use of it helps me and helps me to travel light in life." The eminent philosopher Nigel Warburton (2008) captures this sentiment:

You can really only understand your life if you're prepared to think about your death – the limit of your own existence. Otherwise you're going to spend your days in a dreamy state, acting as if this conscious experience is going to continue forever, ignoring that it has an inevitable end point… So, make the most of your life while you can, and accept that it is going to end.

(p. 49)

Specifically, some interviewees were motivated to leave a positive legacy for future generations. Marilee said, "I want to make the world a better place when I leave it than it was when I arrived… If we all had that view, then the world would be a safer, more equitable place." Marion said, "I've got a grandchild now and I look at him and I wonder about the world that he's growing up in and it makes me feel like I'm trying to do something, to maintain the world that we've got at least, but to help the world." Adrian was particularly concerned not only about his grandchildren but the generations henceforth:

I want my grandkids to have as long a life as possible, and so I'm taking action – That's the major motivation... I do believe in the precautionary principle... thinking about the consequences for your grandchildren's grandchildren's grandchildren, till seven generations. Very much a believer in that now. That's a major ethic that drives me.

Nicole thinks about her child and her wider community. She said:

Every single day it comes back to legacy. My 11-year-old. My reason for being right now... That's what drives me. It's making the world better for our young people, because I know that he's got to live in this world. Showing him what is possible... I have to ensure there's more empowered Indigenous people out there... Because all the other little boys and girls need to see it as well... I'm drawing up on the strength – I'm following in their footsteps – I recognise my ancestors... A pre-ordained beautiful journey my ancestors have set me upon... I'm a Larrakia [Aboriginal Australian from Darwin] person living on Larrakia country. This is my family's Country. I want to make sure again everyone's living their best lives, being their best selves, and Indigenous people are being afforded the same opportunities as everybody else.

Noraini adopts the intergenerational perspective of her father: "There was this mango that my father really wanted. He planted it from a seed and said 'I'm not planting this for me, and perhaps not even for you, it's for my grandkids and my great grandkids, I'm planting this to leave something behind.'" Similarly, Caroline said:

What are we going to leave for the future generations? I've been so lucky to see such a beautiful planet so far – what can we do to help it?... I want to keep it like that not destroying what we have – at least we can keep it as nice as it is.

This theme can be linked to our evolutionary pasts where deep empathetic interest in other humans beyond our immediate family have been shown to be fundamental to the development of human beings as a species (Blaffer Hrdy, 2011). Caring for others, connecting to community, being aware of the interdependence of all things, understanding our place in the universe, and seeing the value of cooperation have been vital evolutionary survival tactics:

Love, sympathy and self-sacrifice certainly play an immense part in the progressive development of our moral feelings. But it is not love and not even sympathy upon which Society is based in [hu] mankind. It is the conscience – be it only at the stage of an instinct – of human solidarity. It is the unconscious recognition of the force that is borrowed by each [person] from the practice of mutual aid; of the close dependency of everyone's happiness upon the happiness of all; and of the sense of justice, or equity, which brings the individual to consider the rights of every other individual as equal to his own.

(Kropotkin, 2012, p. 6)

Reflection

Activists are often asked by friends, relatives, and other interested parties why they do what they do, so being able to articulate a response will be beneficial for understanding yourself on a deeper level and may inspire others. When Doctor Helen Caldicott, campaigner against nuclear weapons, was asked why she does what she does, she replied:

You have to look in the mirror every morning and say to yourself: "What am I doing today to save life on the planet?" There can be nothing else more important... I've spent the last 46 years deeply obsessed by the fact that our lives are threatened every day. I'm a paediatrician, a doctor, who took the Hippocratic oath promising to save lives. And that's what I've been trying to do for 46 years. I worked out my own path and what I had to do. You must do the same. You can write letters to the paper and use social media to the max. You can join things like Greenpeace. I've had eight death threats. I was a terrible threat to the military industrial complex. But I thought "What is my life compared to life on the planet?" I'm prepared to die for this. Martin Luther King once said, "If a man hasn't discovered something that he will die for, he isn't fit to live." And I would support that.

(Caldicott & Boag, 2008, p. 77)

When activists reflect upon the reasons for why they do what they do, and develop the skills to articulate this clearly, it can be a significant step in their journey of personal development. The process may be a catalyst for finding out whether activism helps make life more meaningful.

Psychiatrist, Holocaust survivor, and writer, Viktor E. Frankl (1946/ 2013) theorised that the meaning of life cannot be found searching for pleasure or power. He extended the theories of Austrian neurologist Sigmund Freud, and the German philosopher Friedrich Nietzsche. Frankl believed that the meaning of life is a quest to find meaning and suggests that everyone needs *something* to live for. It involves confronting what life has to offer and taking responsibility to find out how tasks will be fulfilling. This approach is person dependent, context dependent, historically dependent, and place-based. This quotation comes from the second part of the book after Frankl describes the characteristics of the people who survived the concentration camps:

> There is nothing in the world, I venture to say, that would so effectively help one to survive even the worst conditions as the knowledge that there is a meaning in one's life. There is much wisdom in the words of Nietzsche: "He who has a why to live for can bear almost any how."
>
> (Frankl, 2013, p. 140)

Frankl (2013) suggests that this meaning can be found in three different sources: work, love, and courage. Work is about actively contributing such as creating or doing a deed; love is experiencing something or caring for someone or something else; and courage is about our attitude when suffering.

Philosopher Susan Wolf (2010) proposes a way of thinking about human motivation that can provide guidance for us to find our own meaning: "A meaningful life is a life that a.) the subject finds fulfilling, and b.) contributes to or connects positively with something the value of which has its source outside the subject" (p. 20). Meaningfulness exists when a person can find something they can do and enjoy doing, and these efforts are applied to a project that is worthwhile. Sometimes projects that are worthwhile are referred to as *larger than oneself*, but it is not about size. The point is that the person is actively applying themselves to something that is outside of their self. Thinking about whether our actions contribute to something beyond our immediate needs and intersect with something beyond ourselves helps us connect to something bigger and feel part of the grand scheme of things. This approach does not discern whether the project produces positive or negative outcomes (Chapter Seven on evaluation covers this topic), but it does provide a first step in understanding whether our motivations and actions will help us find fulfilment.

Wolf (2010) notes that when people find a passion and pursue something they care about this can result in an engaging and even thrilling feeling of joy. Paradoxically, she also notes that this state may be accompanied by stress and suffering. Wolf (2010) suggests, "Excitement is compatible with fear; love is compatible with sadness; the activities one finds fulfilling are perhaps more often than not difficult and demanding" (p. 112). She encourages us to be cautious and humble when asking ourselves questions about whether we find our activities both fulfilling and contributing to something beyond ourselves. Wolf (2010) suggests that so many valuable effects (social capital, community spirit, resilience, and creativity) can arise from actions associated with sport, art, knowledge creation, social, environmental, or political causes.

Activism, predominantly because of the focus on working towards changing things that are beyond oneself, lends itself to being a meaningful activity. Strong evidence exists about the personal benefits of working on projects that align with our values and skills, which we find meaningful, satisfying, and fulfilling (Soren & Ryff, 2023). Fulfilment will be explored in-depth in the subsequent chapter. As a starting point, writing answers to these reflective questions may help identify whether there is a connection between your motivation, whether your actions have value beyond yourself, and what you find fulfilling.

Motivation

- What motivates you into activism?
- Are there multiple motivations that intersect?
- How have your motivations changed over time?
- Which actions are driven by an intrinsic passion?
- What would success look like for you and how important is it that you achieve success?

Application to something larger than oneself

- How are you creating, promoting, protecting, honouring, preserving, affirming, adding something?
- In what ways are the object of your attention worthy of your skills, passion, devotion, and time?
- Can you identify where your actions have an affect beyond your self-interest?
- How does your activism align with your moral code, or religious worldview, or something else?

Fulfilment

- How does activism make you feel?
- What do you enjoy doing?

- What activities are you passionately engaged with or find gripping?
- What sacrifices do you have to make to engage in activism?
- What experience, skills, and attributes can you offer?

Conclusion

This chapter encourages activists to think deeply about why they are doing what they do. It explored the motivations, values, and beliefs of everyday activists to understand why they want contribute to change. Understanding human motivation can be a way of elucidating the elements of a meaningful life. The people interviewed for this book were nominated because they have been actively applying their skills and using their talents to engage in a variety of different activities. These individuals are motivated by a myriad of reasons. However, the common thread in interviewees' discussions on motivation was that they were working on projects that were concerned with issues beyond themselves as individuals. The philosopher Peter Singer (1993) suggests how valuable reflection and introspection are for developing "practical wisdom" that can help us determine what is important in life. Singer (1993) summarises a fundamental lesson that he formulated after ruminating on the work of Frankl and other philosophers: "[Humans have a] need for commitment to a cause larger than the self, if we are to find genuine self-esteem, and to be all we can be" (p. 255).

Note

1 Also transcribed as *sevā,* meaning selfless service in South Asian cultures, particularly Sikhism and Hinduism

References

Ágoston, C., Csaba, B., Nagy, B., Kőváry, Z., Dúll, A., Rácz, J., & Demetrovics, Z. (2022). Identifying types of eco-anxiety, eco-guilt, eco-grief, and eco-coping in a climate-sensitive population: A qualitative study. *International Journal of Environmental Research and Public Health*, 19(4), 2461–2470. doi: https://doi.org/10.3390/ijerph19042461

Allison, E. A. (2015). The spiritual significance of glaciers in an age of climate change. *WIREs Climate Change*, 6(5), 493–508. doi: https://doi.org/10.1002/wcc.354

Antonetti, P., & Maklan, S. (2014). Feelings that make a difference: How guilt and pride convince consumers of the effectiveness of sustainable consumption choices. *Journal of Business Ethics*, 124(1), 117–134. doi: https://doi.org/10.1007/s10551-013-1841-9

Atkinson, J. (2002). *Trauma trails, recreating song lines: The transgenerational effects of trauma in Indigenous Australia*. Spinifex Press.

Blaffer Hrdy, S. (2011). *Mothers and others: The evolutionary origins of mutual understanding*. Harvard University Press.

Boon, S. (2021). Look past the woods – each tree is an individual to be cherished. *Psyche*. https://psyche.co/ideas/look-past-the-woods-each-tree-is-a n-individual-to-be-cherished.

Buschmeier, M., & von Kellenbach, K. (2022). Introduction: Guilt as a force of cultural transformation. In *Guilt* (pp. 1–26). Oxford University Press. doi: https://doi.org/10.1093/oso/9780197557433.003.0001

Caldicott, H., & Boag, Z. (2008). The end of life on earth. *New Philosopher*, 19, 72–77.

Culiberg, B., Cho, H., Kos Koklic, M., & Zabkar, V. (2023). The role of moral foundations, anticipated guilt and personal responsibility in predicting anti-consumption for environmental reasons. *Journal of Business Ethics*, 182(2), 465–481. doi: https://doi.org/10.1007/S10551-021-05016-7/FIGURES/1

Eze, M. (2011). I am because you are. *The UNESCO Courier*. https://en. unesco.org/courier/octobre-decembre-2011/i-am-because-you-are

Foster, C. (2022). Stand up for those without a voice. *The Big Issue*, 675(Nov), 18–19.

Frankl, V. E. (2013). *Man's search for meaning: The classic tribute to hope from the Holocaust*. Ebury Publishing.

Halifax, J. (2018). *Standing at the edge: Finding freedom where fear and courage meet*. MacMillan Publishers.

Harðardóttir, J. (2021). *Sacred landscapes: Solastalgia and spirituality in a melting world*. Radio National, ABC. https://www.abc.net.au/radionationa l/programs/soul-search/sacred-landscapes:-solastalgia-and-spirituality-in-a -melting-wo/13585792

Kropotkin, P. (2012). *Mutual aid: A factor of evolution*. Dover Publications (Original published in 1902).

Land, C. (2015). *Decolonizing solidarity – Dilemmas and directions for supporters of Indigenous struggles*. Zed Books.

Lindenmeier, J., Lwin, M., Andersch, H., Phau, I., & Seemann, A.-K. (2017). Anticipated consumer guilt: an investigation into its antecedents and consequences for fair-trade consumption. *Journal of Macromarketing*, 37(4), 444–459. doi: https://doi.org/10.1177/0276146717723964

Ma'anit, A. (2007). Guilt complex. *New Internationalist*, 406. https://newint. org/features/special/2007/11/01/special_feature

Marin, L. (2021). Sharing (mis) information on social networking sites. An exploration of the norms for distributing content authored by others. *Ethics and Information Technology*, 23 (3), 363–372. doi: https://doi.org/10.1007/ s10676-021-09578-y

Maté, G., & Maté, D. (2022). *The myth of normal – Trauma, illness, and healing in a toxic culture*. Vermilion.

McKegg, K. (2019). White privilege and the decolonization work needed in evaluation to support Indigenous sovereignty and self-determination. *Canadian Journal of Program Evaluation*, 34 (2), 357–367. doi: https://doi.org/10.3138/cjpe.67978.

Motta, G. (2018). *Dynamics and policies of prejudice from the eighteenth to the twenty-first century.* Cambridge Scholars Publishing.

Peace News. (2009). Activism and... guilt. *For Nonviolent Revolution: Peace News.* https://peacenews.info/node/5811/activism-and-guilt

Rozelle-Stone, A., & Davis, Benjamin P. (2023). *Simone Weil: The Stanford Encyclopedia of Philosophy.* https://plato.stanford.edu/archives/sum2023/entries/simone-weil/

Seth, A. (2021). *Being you: A new science of consciousness.* Faber.

Singer, P. (1993). *How are we to live.* Random House Australia.

Singer, P. (2009). *Animal liberation: The definitive classic of the animal movement.* Harper Perennial.

Smith, L. T. (2012). *Decolonizing methodologies* (2nd Edn.). Zed Books.

Soren, A., & Ryff, C. D. (2023). Meaningful work, well-being, and health: Enacting a eudaimonic vision. *International Journal of Environmental Research and Public Health*, 20 (16), 6570. doi: https://doi.org/10.3390/ijerph20166570

Utt, J. (2014). True solidarity: Moving past privilege guilt. *Everyday Feminism.* https://everydayfeminism.com/2014/03/moving-past-privilege-guilt/.

Warburton. (2008). Thinking about your death. *New Philosopher*, 19, 49–50.

Watson, D. (2022). The king and us: How should we think about Australia's new monarch, King Charles III? *The Monthly*, November.

Wolf, S. (2010). *Meaning in life and why it matters.* Princeton University Press.

Zaki, J. (2020). *The war for kindness: Building empathy in a fractured world.* Broadway Books.

4 Finding fulfilment

This chapter on *finding fulfilment* unpacks how activism can enhance personal growth and self-development. The previous chapter explored underlying motivations and proposed that, no matter where the drive was initially sparked, a meaningful life can be found when activists apply themselves to something outside of themself and the activities are fulfilling. It ended with some questions about motivation and satisfaction, encouraging readers to consider what they enjoy doing, what they find gripping, what sacrifices they make, and what skills and attributes are needed. This chapter moves beyond motivation and satisfaction to unpack the concept of personal fulfilment and assess how activism can be a catalyst for personal development.

Activism resulted in interviewees experiencing a wide range of sensations that were not mutually exclusive, were sometimes conflicting, and could often be felt within seconds of each other. Emotions varied depending upon the circumstances and could easily change in response to their thoughts or external factors. Adverse outcomes included fatigue, exhaustion, and personal sacrifice. Many grappled with self-doubt and perceived inadequacy. Many felt a sense of purpose, utility, euphoria, and gratification. Interviewees also talked about social connection and the social benefits that come from developing a profound connection with others whilst collaboratively working towards a common goal. This was underscored as a notable advantage of interviewees' involvement in activist endeavours.

Certain interviewees have adeptly balanced these emotional extremes, steering towards self-actualisation. By harnessing their emotional reactions as springboards for introspection and growth, they have refined their core beliefs, assessed their values, questioned their biases, re-evaluated accepted narratives, and recognised what energises or drains them. Through these reflections, they exemplify how activism can serve as a fulcrum for personal evolution. Findings from the

DOI: 10.4324/9781003333982-4

interviewees adds to research evidence which shows that people who define their own sense of purpose are more likely to live a healthier and longer life (Soren & Ryff, 2023). In this chapter we emphasise the value of developing self-awareness by finding a balance between the positive effects of activism and critical self-reflection. We argue that activism can help us be the best version of ourselves and provide a path towards self-actualisation. The chapter concludes with thought-provoking questions aimed at encouraging readers to introspect on the role of activism in their journey towards realising their full potential.

Adverse outcomes

Some of the interviewees in this book make significant sacrifices for their cause. They place their lives in danger, risk the safety of their family members, have been arrested or risk arrest for their activism, have been tortured, forced to flee, experience social ostracism, and place financial and emotional strain on their families. Even going against family expectations takes a personal toll. Serena said, "I think for me the hardest sacrifice was sort of going, 'If I'm not a lawyer, then what am I?'" Additional time-consuming activities beyond their normal roles and duties were also reported as being stressful and tiring. Putima finds it draining when she cannot stop thinking about her activities, "You don't realise it until later when you're just so exhausted." Nafiz spoke of feeling physically exhausted, "Working for people, and for betterment of people, yes, physically it can be sometimes rigorous and tiring."

Some interviewees simply accept that activism is something they must do as part of their duty, which comes with personal discomfort and potential negative consequences. However, they outline that the collective benefits outweigh any personal toll. Ekawati notes that in Indonesia: "many activists in my circle tend to adopt 'collective benefits outweigh personal suffering'". Similarly, Delphine understood that making sacrifices was part of the process: "I'm doing it because it's the right thing to do. I might not be happy with that but that's not the point… When I do something for someone, I'm doing it because it has to be done. Not because of how they might make me feel after it."

Serving others and undertaking activities such as promoting social justice, being pro-environmental, and challenging social inequalities require the investment of high levels of energy and resources over the long-term. This type of activism can deplete the resources of the activist. References to self-sacrifice can be found in many historical and cultural sources about activists. To encourage people to take up the challenge,

Martin Luther King, Jr. (1956/2010), said, "There is nothing more majestic than the determined courage of individuals willing to suffer and sacrifice for their freedom and dignity" (p. 1). Researchers have found that people were more willing to undertake self-sacrificing behaviours when their actions align with their social values (Jasko et al., 2019). However, as covered in more depth in Chapter Eight on self-care, researchers have also noted that experiencing burnout, acute psychological distress, depression, trauma, stress, and exhaustion can be some of the demoralising effects associated with activism (Wang, 2019).

Even if some interviewees mentioned feeling a positive emotion such as a flash of elation, they indicated that feeling good is regularly followed by other intense, and often negative, emotions. Some interviewees indicated that feeling *good* was a sign that they were self-absorbed or righteous. They did not like this hint of arrogance and ego. Mike articulated the risk in this way:

> When I was filled with the hubris of youth, [contributing] tended to make me feel pretty good about myself, nothing wrong with that, except it can lead to feelings of virtuous righteousness, and even a kind of judgemental superiority, which is a bit like climbing a set of stairs at the top of which is a terrifying drop!

Although Marilee does not necessarily think that feeling good about yourself is problematic, as she is always finding ways to extend her own and others' self-esteem, she also expressed cautiousness: "There's a moral high horse to living in line with your values. There's a morality to it that you can feel really good about and maybe even a bit smug at times." Jody acknowledged she did feel good when she achieved a goal, but very quickly clarified with: "I want to do it for the right purposes."

Other interviewees reported that after feeling *good* they instantly thought about the insignificance of their efforts. Alicia said, "I have a very short moment of where I'm quite happy and jubilant, and then the realisation of 'There's so much more to do and that I need to deliver', it just overtakes." Roy acknowledged that activism made him feel good but quickly provided a qualification, "I'm okay, but not satisfied yet. I feel like there's more I could be doing." For Alyssa, the larger looming issues continue to be a burden to carry: "Nothing ever really feels like a win. I have the knowledge that I'm contributing to a broader regime of change and chipping away at a giant edifice. The wins usually come with a caveat... it doesn't really feel like success."

Some interviewees questioned whether they were doing the right thing, whether they were the right person, and whether someone else would be more effective. Nigel said, "There's a very strong element of feeling like I know what needs to be done better in the advocacy space, but not having the skillset to do it." Ben acknowledged that he did treasure moments of success, but he added:

> If I'm honest, there's been many times in the last 10–12 years of working in this space where I feel the imposter syndrome... like that what I'm doing isn't really making that much of a difference. Do I think that this is an important enough issue to be focusing my time and my career and my energy on? When I could be working on refugee rights or climate change?... I think there's definitely a feeling always that I'm not doing enough or I'm not doing the right things... Sometimes I try and do more, and sometimes it sits in the back of my mind subconsciously... a feeling of nascent guilt that never goes away.

Feeling like an imposter can be a debilitating or a constructive emotional response depending upon how it is handled. Research defines *imposter syndrome* as "high-achieving individuals who, despite their objective successes, fail to internalize their accomplishments and have persistent self-doubt and fear of being exposed as a fraud or impostor" (Bravata et al., 2020, p. 1252). Although it may not be useful to spend too much time thinking of ourselves as imposters, these feelings can be reframed as opportunities for constructive critical self-reflection. If a person does not contemplate where they need to improve, then they may be overly sure of themselves; those who express doubt may actually be applying an appropriate level of self-scrutiny (Tavris & Aronson, 2007).

Favourable outcomes

While not always the case, undertaking activist activities can provide a sense of satisfaction and purpose. Jenny shared that helping others, "helps life feel purposeful and meaningful." Dia said, "It makes me feel complete, and I want to do more and more, it makes me feel purposeful." Domuto also used the word *purpose* when he shared the effects of contributing: "It's important because [peacebuilding] work can make us feel good about ourselves and give us a sense of purpose, it's important to protect and improve our mental health and well-being." Talia said, "It makes me feel good. As simple as that sounds, it's good to feel like I'm playing my part." This was expressed by Iram who humbly reflected on the significant impact she makes: "I may not see lot of change out of my efforts but do see small ones here and there and that make me feel good. In my small ways whatever I can contribute makes me feel better." Marion shared how volunteering gave her purpose after retirement:

> All my life before retirement, meaning had come from work and relationships... what you contribute was very work focused and I loved work. I was very fortunate to work in all the challenging jobs that I really enjoyed doing. Meaning after retirement is coming from feeling unity and giving back – offering something to society – you've got an important role to play like everyone else.

In 1943, psychologist Abraham Maslow presented a theory of psychological health in the form of a hierarchy of needs. He places basic needs at the bottom and a personal development need, that is about fulfilling our potential as a human being through seeking meaning, at the pinnacle. It is sometimes assumed that Maslow's (1943) hierarchy of needs dictates that people's basic needs must be met before they can support others through activism. After interviewing several young men living in the muddy, unsafe, and bleak refugee camps of Cox's Bazar, Bangladesh, we contend that this is not the case and activists exist in all settings, including those where activism is a fight for survival. Arif feels the weight of responsibility toward his fellow Rohingya and deeply understands the struggles and lack of rights facing his people. He explained: "we have been surviving refugee life since 2017 with no freedom to move, no quality educational access to study, and no hope to return to our Motherland with our full rights and justice." Despite crushing hopelessness that permeates the situation facing his people, he continues to advocate for his community and encourage others to join him.

Contributing becomes intrinsic to activists' sense of self through developing a personal identity around their activist role. Alicia finds that contributing "gives me a sense of worth, just makes me feel good, and I respect myself, and that's the most important thing to me, is having respect for myself." Alyssa said, "I feel like everything I do defines me. So if I wasn't doing this stuff, I wouldn't know who I am. I feel like it legitimises me. It gives me a justification for being around, that I'm trying to do something good." Mehreen relates to Mahatma Gandhi's understanding of how the act of contributing helps to find out more about herself:

> I think it was Gandhi who said losing yourself in the service of others helps you figure out who you are and what makes you happy. It doesn't take much and it's not about money; it's just

about time. Not about solving their problem. Just giving your ear and your time... I have two ears, I can listen, I can spend time with you, I can make you feel a little bit better than when we met. It feels good for sure. Doing whatever you can with what you have.

Mehreen is making the point that discovering your own purpose can come through the process of contributing when the needs of something or someone else are placed above your own. This was also expressed by a character in Chekhov's (2009) short story *The wife* where the protagonist begs her husband to let her contribute to a local committee aiming to address poverty:

> "Pavel Andreitch," she said, smiling mournfully... "Call this" – she pointed to her papers – "self-deception, feminine logic, a mistake, as you like; But do not hinder me. It's all that is left me in life." She turned away and paused. "Before this I had nothing. I have wasted my youth in fighting with you. Now I have caught this and I am living; I'm happy... It seems to me that I have found in this a means of justifying my existence."
>
> (p. 19)

A proportion of the interviewees described an emotional *buzz* from contributing. They described feeling great, happy, positive, joyful, and jubilant. When Adrian works on a stall, he recounted that, "I get a huge charge and good feelings out of that." Kim recalled the celebrations when her group successfully prevented a quarry expansion: "I'm remembering that cake we had with all the gravel lollies on it for the quarry where we were like 'We won!' I mean that was a real huge buzz. What a buzz!" Nigel similarly said, "It makes me feel pretty good actually. It's probably what I get out of life more than anything else." Roy said, "There's nothing more fun than anonymously giving someone money. If you want a real rush, put some money in an envelope and put it in someone's mailbox. I love it, it's what I live for. It's the same stuff that washes your brain when you take drugs." Nicole said, "It fills my cup, it really does. I don't do it for the accolades... It's just very organic... It makes me feel happy... I'm happy inside. I'm content. I'm actively contributing. I love volunteering my time." Delphine had this response to the happiness of a lady she was supporting: "I was just so happy that this person was able to get some joy out of her life in that moment... I think that was beautiful... I don't know if it was a selfish thing, but I certainly was happy to see her sing." Despite the appalling circumstances and deeply hopeless situation facing Rohingya refugees in the Bangladeshi camps, Arif expressed:

It makes me [feel] strong, it makes me move forward to save others. I get very happy and strongly encouraged when I can do something good for someone. I don't and never will give up. I keep moving forward [even] if I fail, no matter what it takes and how difficult.

Similarly, even as Jamal risks his life photo-journalling the situation of his fellow Rohingya in the refugee camp, he said:

I love my work. I love my job very much. I let the whole world know what is happening in every moment of my nation. So, at the cost of my life... I will continue to work for my nation. I have a lot of respect for this work of mine. I am very proud of my work.

This sensation was described in an article titled "Inside the sausage factory" (Turner, 2022). An advocate at the COP26 summit said that "It felt like Christmas" when they advocated to tweak a few words on a resolution. Even though they would never know if it made an impact, he said, "Just the chance, though, made it the most worthwhile two weeks of my life" (cited in Turner, 2022, p. 6).

The interviewees were energised by their work and felt an emotional boost when they achieved their goals or were witness to the impact. Nafiz coincidentally heard about how his training affected peoples' lives in a positive way and noted, "You cannot put any value to it! Any project or any training is contributing to someone not only in their workplace but also in their personal life – that is really great." Although Khalil humbly suggested that his facilitating role in bringing a panel of young people together was minor, he did experience a profound emotional reaction: "All of them are really amazing, writing about important things, and I feel beautiful satisfaction. It made me think that all the work I do isn't for no reason." Mosrat described how she felt when her program became self-sustaining and did not require further personal input:

It feels really very good, when I feel that they actually own the program – Always I tell people this, you should not join in such kind of work where actually you can only earn money... You should always choose the path, where you feel your attraction and affection – you'll have lots to work with.

Social connection

Most interviewees gave examples of establishing strong relationships, finding a sense of collegiality, and enjoying the feeling of being part of a like-minded community. Activism provides unique opportunities to meet new people with similar values and motivations and to share and exchange ideas beyond their immediate social circle. It can facilitate a sense of belonging, transformation, and connectedness, as well as increased levels of personal happiness, social skills, compassion, health, wellbeing, and cognitive ability (Gray & Stevenson, 2020; Haddock, 2015; Soren & Ryff, 2023). Interviewees said they relished the joint celebrations where members of their groups reflect together on their achievements. They highly valued the sense of team spirit and solidarity they experience when working with others trying to make a difference to a common cause.

Nigel said that he values the opportunity of trying to make a difference alongside likeminded people with whom he shared a "sense of solidarity". Elise said she enjoys the social dynamics when celebrating wins, "It's the sense of community and collaboration you get out of working with likeminded people... it's really interesting... just that sense of like mindedness... the sense of collegiality... being a part of a coalition [provides] positive reinforcement of being part of a bigger movement." Noraini values the camaraderie: "I was lucky enough to have [the organisation] which was a place I could get to every single day – helping other people and even being around my colleagues – was really good." Caroline shared, "When you do some actions like planting a garden at the museum, or cleaning the beach and doing it with more people, first it creates a good spirit with the team you're working with and it's really fantastic." Vijeta suggested that making interpersonal connections was a part of her personal development: "When you're connecting more with people in a deeper way, you grow older better, that's my aim in life. You start like a raw fruit and then as you grow you become ripe." Alex said that her community garden, "makes me feel like a really valued member of my community."

Seeing activism as an opportunity to connect with others can be an enriching foundation upon which to base effective actions. Lisa Yun Lee (2009) in *Counterpoints – Queers organizing for public education and justice*, writes about the demanding, infuriating, and difficult role of an activist, but implores us to ensure it is also pleasurable. Lee (2009) states, "Activism that does not inventively engage our sentient and sensual selves and does not give us joy or make us happy, is unsustainable and also unattractive" (p. 105). She encourages activists to include existing friends and family and make new friends so as to take pleasure in eating, laughing, reading, creating, imagining, sharing, and singing together. She believes that it is the elements of camaraderie that can sustain groups of reformers over the long-term (Lee, 2009). Ultimately, social connection also provides us with opportunities to mentor and be mentored. It can help us to be more inclusive and develop skills in supporting others and thereby be an important step on our journey of self-development.

Finding balance

Becoming aware of the range of emotional reactions related to activism may be the first step in working out how to find an effective balance. Finding balance creates the conditions for enabling personal development to occur as part of a long-term ongoing process. Khalil provided an example of how he balances the joy of celebrating whilst recognising his small place in the big picture:

> I see myself as a small contributor... but part of the success is mine... I contributed. And I do believe it's important to celebrate as activists or agents of change... Ok there's injustice in the world – but you did something about it – You need to celebrate! Enjoy small satisfactions! Make activism and advocacy a beautiful thing so people want to join you.

Jenny contributed a balanced perspective, "Giving to others gives to you; it's so enriching. You shouldn't just serve others. I follow the idea of it being mutually beneficial." Dia covered the gamut of emotions:

> It makes me feel complete, and I want to do more and more, it makes me feel purposeful. I always believe *lead by example*, and that has helped my child, to see the way I'm leading my life. I'm always reflecting my core values through what I do. Empathy is a privilege and a responsibility, not something you do to feel good. Vulnerability has helped me to appreciate things, it's not a weakness, it's a growth opportunity. Making a contribution radiates to other people... It's the value creation, your legacy. What did you do today, for you or for what you love doing? Keep adding to that jar of value creation. And then that's a ripple effect that impacts your home, and your community. Every day wake up with focus, and it has to be your focus, and that is part of your self-love.

Finding balance can often be about gaining perspective and coming to terms with what your role can realistically achieve. Philosopher Peter Singer (2016) in his book *The most good you can do* suggests demanding no more of yourself than you can give whilst remaining cheerful. Buddhism (as well as other philosophies and religions) outlines pathways to balance caring for others with self-care. Lama Surya Das (2004) in the book titled *Letting go of the person you used to be*, explains:

Inner detachment, remember, is not synonymous with indifference. We do still care about others, as well as social justice issues and concerns, but we are far less invested in desirable outcomes. Inner balance and equanimity help us feel clearer and less volatile about what we are doing and why. We can flow better and roll more gracefully with the rollicking punches and weaving bumpy road-way of life... Doing the best you can, here and now, continues to be the Buddhist way. Letting go means letting come and go – letting be. It means coming to accept what can't be changed even while working for positive growth, change, and transformation.

(p. 86)

Reflection

Maslow's (1943) hierarchy of needs and the concepts of self-actualisation and self-transcendence can provide us with some useful tools. These concepts have significant historical and recent empirical support for understanding human motivation, personality, and the characteristics of self-actualisation (Kaufman, 2023). In the *Encyclopedia of quality of life and well-being research*, self-actualisation is defined as "a motive, striving, need, or goal involving behaviours directed toward the fulfillment of personal potentials. Its meaning is captured in the phrase 'becoming the best you can be'" (Waterman, 2014, p. 5743). Having the tendency to actualise or progress towards finding fulfilment is based on an understanding that individuals on a biological, psychological, and social level, under whatever circumstances, will be drawn towards developing their best selves. It is an innate feeling or intuition that can help us find the optimal expression of who we naturally are. There is no end point, but self-actualisation can instead be understood as an ongoing journey or process that will adjust and change over our lifetime (Worth, 2017).

When the interviewees for this book described finding their purpose and using their skills, creativity, awareness, and courage to contribute to something beyond themselves, their reactions align with the concept of self-actualisation. Kaufman (2023) found that people who said they experienced self-actualisation were motivated by personal growth, exploration, and a connection with humanity and nature. Self-actualisation has been found to correlate with wellbeing factors such as self-acceptance, meaningful interpersonal relationships, feeling in control, independence, and purpose. Evidence drawn from the interviewees indicate that only the bare necessities are required to be satisfied before individuals can start thinking about self-actualisation and self-transcendence. Interviewees living in a refugee camp with only tarpaulins for shelter provided examples of how activists in extreme circumstances think beyond themselves and apply their skills to a higher purpose. This is supported by research which showed that people in difficult circumstances can experience benefits from helping others and using their personal strengths (Tweed et al., 2017).

Self-transcendence is particularly important for activists who are thinking beyond themselves and outwardly towards the communities with whom they are working; it is about acting and working in ways that are in service of others. Related to the message purported in Chapter Three on underlying motivations, Wong (2016) suggests self-transcendence is about applying ourselves to our cause and enjoying feeling absorbed by something beyond ourselves. Worth and Smith (2021) argue that attempting to reach this state is part of our spiritual nature and can influence our healing and wellbeing; we find out more about ourselves on a deeper level and feel more fulfilled and actualised as a human being.

The path to fulfilling potential involves taking the time to understand where we feel the most burden and where we feel the most reward. Making personal sacrifices in the service of others may result in negative consequences and adverse outcomes. Seeing others as people who need your *help* or a situation that needs an intervention from you to *fix it* may mean you expect or need them to express gratitude. Or you may become dependent on them so that you can feel valued. Therefore, it is not about seeing others as needing our help, but instead being willing to guide a process of change by sharing the best parts of ourselves in an open and resourceful way for a mutually beneficial outcome (Worth, 2017).

We also need to listen to our inner warning bells and self-doubt. Incorporating a rational critical reflection process can elucidate what needs to change. Our inner negative alarm system may be telling us that we lack the right skills, we have taken on too much, or we need to change our approach. These gut feelings, whether they be negative or even the positive warm glowing buzz variety, should not necessarily be ignored – they all provide valuable insight into where we should be placing our efforts. Critical reflection can liberate us from actions that sap our vitality and help us find energy.

Recognising that different situations conjure a complex array of wide-ranging emotional responses, it may be beneficial to identify some of the contextual factors such as time and place where the effects are felt so they can be interpreted with some clarity. These reflective questions aim to assist with developing self-awareness and identifying a path towards the liberating state of self-actualisation:

- How does my activism work contribute to my long-term goals and aspirations?

- To what extent do I feel a sense of fulfilment and purpose through my activism?
- What steps can I take to align my activism more closely with my self-actualisation goals?
- How does my activism work resonate with my authentic self?
- In what ways has my activism work evolved over time?
- What skills and knowledge have I gained through my activism?
- What challenges have I faced in my activism journey and how have they shaped my resilience and problem-solving skills?
- What have I learned from collaborating with others in activism?

Conclusion

This chapter presents the increasingly robust evidence that people with a sense of purpose, such as activists, are likely to be healthier and live longer (Ryff, 2018; Soren & Ryff, 2023). Using our strengths and sharing what we have to offer in the service of others can bring a deep joy and sense of stability (Tweed et al., 2017). Reflecting on the personal effects that arise from activism can help you know more about yourself and consider what personal liberation and fulfilment would look like. Understanding how activism makes you feel can help you understand more about why you are an activist and whether you are working towards fulfilling your potential as a self-actualised human being. As Australian children's author Andy Griffiths (2004) encourages:

> I believe it is never too early – or late – to start asking the questions: "What is it that really absorbs me?" And "What am I uniquely suited to being able to contribute to both my life and the lives of others?" – and then to have the courage, patience, and persistence to act on the answers as if your life depends on it. Because, in a very real sense, it does.
>
> (p. 121)

The process of developing self-awareness will be different for every person. However, a valuable initial step involves identifying when we are energised and flourishing or when a sense of self-doubt is reminding us to critically self-reflect. When is it appropriate for us as heavily flawed hypocritical individuals to feel proud of our work? What do we need to do to uncover those flaws and hypocritical characteristics? How can we actively work to rectify them little by little whilst maintaining self-compassion? Reaching self-actualisation or self-transcendence whilst remaining humble, reflective, curious, and mindful of the consequences of our actions can assist with finding the balance between personal fulfilment and effective activism.

References

Bravata, D. M., Watts, S. A., Keefer, A. L., Madhusudhan, D. K., Taylor, K. T., Clark, D. M., Nelson, R. S., Cokley, K. O., & Hagg, H. K. (2020). Prevalence, predictors, and treatment of impostor syndrome: A systematic review. *Journal of General Internal Medicine*, 35(4), 1252–1275. doi: https://doi.org/10.1007/s11606-019-05364-1

Chekhov, A. (2009). The wife. In *Plays by Anton Chekhov* (EBook #7986). Project Gutenberg.

Gray, D., & Stevenson, C. (2020). How can 'we' help? Exploring the role of shared social identity in the experiences and benefits of volunteering. *Journal of Community & Applied Social Psychology*, 30(4), 341–353. doi: https://doi.org/10.1002/CASP.2448

Griffiths, A. (2004). Andy Griffiths. In J. Marsdon (Ed.), *I believe this: 100 eminent Australians face life's biggest question*. Random House Australia.

Haddock, M. (2015). *UN "Human development report" and OECD's "How's life?" emphasize contribution of volunteering*. Centre for Civil Society Studies Archive: John Hopkins University. http://ccss.jhu.edu/oecd-un-volunteering/

Jasko, K., Szastok, M., Grzymala-Moszczynska, J., Maj, M., & Kruglanski, A. W. (2019). Rebel with a cause: Personal significance from political activism predicts willingness to self-sacrifice. *Journal of Social Issues*, 75(1), 314–349. doi: https://doi.org/10.1111/JOSI.12307

Kaufman, S. B. (2023). Self-actualizing people in the 21st century: Integration with contemporary theory and research on personality and well-being. *Journal of Humanistic Psychology*, 63(1), 51–83. doi: https://doi.org/10.1177/0022167818809187

King, M. (2010). *Stride toward freedom: The Montgomery Story*. Beacon Press. https://kinginstitute.stanford.edu/publications/autobiography-martin-luther-king-jr-chapter-7-montgomery-movement-begins [Original quotation in 1956]

Lee, L. Y. (2009). Afterword: On the pleasures of activism. *Counterpoints*, 340, 105–108. https://www.jstor.org/stable/42979259

Maslow, A. H. (1943). A theory of human motivation. *Psychological Review*, 50(4), 370–396. doi: https://doi.org/10.1037/h0054346

Ryff, C. D. (2018). Eudaimonic well-being. In *Diversity in Harmony – Insights from Psychology* (pp. 375–395). Wiley. doi: https://doi.org/10.1002/9781119362081.ch20

Singer, P. (2016). *The most good you can do: How effective altruism is changing ideas about living ethically*. Text Publishing.

Soren, A., & Ryff, C. D. (2023). Meaningful work, well-being, and health: Enacting a eudaimonic vision. *International Journal of Environmental*

Research and Public Health, 20 (16), 6570. doi: https://doi.org/10.3390/ijerp
h20166570

Surya Das, L. (2004). *Letting go of the person you used to be: Lessons on
change, loss, and spiritual transformation*. Harmony.

Tavris, C., & Aronson, E. (2007). *Mistakes were made (But not by me): Why
we justify foolish beliefs, bad decisions, and hurtful acts*. Harcourt.

Turner, J. (2022). Inside the sausage factory: Jenny Turner reports from
COP26. *London Review of Books*, 44(1). https://www.lrb.co.uk/the-paper/
v44/n01/jenny-turner/inside-the-sausage-factory

Tweed, R. G., Mah, E., Dobrin, M., Van Poele, R., & Conway, L. G. (2017).
How can positive psychology influence public policy and practice? In *Positive Psychology Interventions in Practice* (pp. 257–271). Springer International Publishing. doi: https://doi.org/10.1007/978-3-319-51787-2_15

Wang, C. (2019). Youth activists power global protest movements but sacrifice
mental health. *The American Prospect*. https://prospect.org/world/youth-acti
vists-power-global-protest-movements-but-sacrifice/

Waterman, A. S. (2014). Self-Actualization. In *Encyclopedia of quality of life
and well-being research* (pp. 5743–5746). Springer Netherlands. https://doi.
org/10.1007/978-94-007-0753-5_2626

Wong, P. T. P. (2016). Meaning-seeking, self-transcendence, and well-being. In
A. Batthyány (Ed.), *Logotherapy and Existential Analysis* (pp. 311–321).
Springer. doi: https://doi.org/10.1007/978-3-319-29424-7_27

Worth, P. (2017). Positive psychology interventions: The first intervention is
our self. In *Positive Psychology Interventions in Practice* (pp. 1–12). Springer
International Publishing. doi: https://doi.org/10.1007/978-3-319-51787-2_1

Worth, P., & Smith, M. D. (2021). Clearing the pathways to self-transcendence.
Frontiers in Psychology, 12. doi: https://doi.org/10.3389/fpsyg.2021.648381

5 Cooperating with others

Every aspect of our lives involves teamwork. It is an essential part of sport, school, and workplaces. Everyday activists need to unite with others, foster effective groups to harness skills, and collaborate to be more impactful. Whether the focus is about mobilising people to focus on an issue or organising people in a way that builds the capacity of individuals to take action, working cohesively as a group is needed to ensure change can happen (Minkler & Wakimoto, 2021). Sometimes, in some groups, something magical happens. It can feel like a powerful current running through the team that makes all things seem possible. Some teamwork can be highly motivating. Many people have felt the synergistic vibe that can occur when a group works in a collaborative way to achieve a common goal. Other times, tensions run high, nothing seems to get off the ground, and the group implodes. Some teams can make a person's energy levels dissipate and many have suffered the consequences of situations where negative dynamics have derailed productivity (Torca et al., 2022).

Effective teamwork does not have to be based on luck. It may feel like good fortune when people are working well together, but social psychology provides a recipe for increasing the likelihood of a team achieving higher productivity. If we are working from a place where we acknowledge our interdependence with the people around us and value the benefits that come from working cooperatively instead of competitively, then research is available that can maximise the effectiveness of the team. Academic Morton Deutsch's (1949) initial studies on social interdependence, cooperation, and competition, have influenced a large volume of high-quality evidence. When teams are high functioning, the people involved are more productive and healthy, they feel psychologically safer, they experience synergy and joy, and have greater potential for upskilling (Deutsch, 2011).

This chapter answers the question, *How can I support and motivate others to get and stay involved?* We begin with a focus on recruiting and

DOI: 10.4324/9781003333982-5

engaging people. Then we discuss the importance of establishing meaningful connections and tailoring interactions. Social psychology and the responses from the interviewees are used to illustrate a recipe for how to intentionally create conditions that support effective teamwork. The final section deals with conflict before we conclude with reflective questions to assist application of the recipe in practice.

Engage

Recruiting others is important because it is through conversation that we can develop a groundswell of people who care enough to act for change. Activists often recognise the complexity of their issue, so initiating conversations with others adds another layer of difficulty. As a starting point, it is worth acknowledging that making friends as adults can be hard (Hronis, 2022). Friendships do not emerge organically; they require people to take responsibility and initiate interactions (Franco, 2020). It can be even more difficult to have conversations on topics that require action and change. The positive is that these skills can be learnt.

Many interviewees were quick to offer strategies for how *not* to engage. Some recommended leaving behind righteous indignation. Janelle suggested activists should be, "A little less judgmental... if people think they're being judged for their choices, you've kind of closed the door on them." Mehreen advised that interactions should not be about "proselytising or cramming it down people's throat!" Some interviewees said to avoid being gloomy, grim, or angry. This was illustrated by Ricki's comment, "You have to work at not bringing in the anger for change – because you get more with honey." Dia reiterated, "Too much of the hard selling doesn't work." Vijeta suggested avoiding overbearing interactions especially during the initial stages, "Saying your piece in a respectful way but not forcing your opinions... Don't ask people to commit anything at first, just come if they can... Not pressuring people to put too many hours in if they can't." Marilee emphasised the need to steer clear of theoretical constructs and instead highlight practical suggestions:

> Unless it also comes with some sort of practical and grounded application that can make it land and stick [it will be ineffective]. You can tell people until you're blue in the face whatever the technical information is... but you have to tell them how or give a suggestion that they can action.

The interviewees' suggestions about what not to do are backed up by research. Sprouting negative messaging such as fear and anger can just exacerbate already reactive situations and potentially inspire more resistance to tackling an issue (Espen Stoknes, 2015; Shenker-Osorio, 2021).

Initiate

One of the most important things we can do is ask people to join us. Talking about the issue in an effective way is a powerful tool. Our family, friends, colleagues, and community are an important source of support and resources waiting to be invited to participate in ways that suit their skills. Simply being asked to join is one of the major predictors of ongoing engagement, especially when the invitation comes from someone who is considered a friend or family (Effective Activist, 2023). The organisation *Climate for Change* operates on the basis that conversation is one the key mechanisms that inspires social change. Their work focuses on helping people initiate conversations because, "People process information, commit to ideas and action when in dialogue with people that they trust" (Climate for Change, 2023, p. 1). Research has also shown across multiple studies that door-to-door canvassing is one of the most effective strategies for engaging people with an issue – the more personal the interaction the greater the impact (Green & Gerber, 2015).

Adrian has a list of talking points that he and his colleagues use when initiating conversations with strangers about climate change. Susanne tailors quantitative and qualitative data into a narrative format for different audiences. Tanya packages anti-horse racing information into non-confrontational formats so that people can distribute it in appropriate ways through their workplaces or social circles. Dia's core group of 15 members are encouraged to reach out to another 20 of their friends, "Mostly it is purely word of mouth. And our donors tell their friends." Matt also shares what he is working on with his local community through social media:

> I try and share what I'm doing. Things like our energy efficient house, it's probably the most energy efficient house in our neighbourhood. I try and promote some of that through Instagram to explain how poorly the average house is built and how with a little bit of effort it can be way better. If community members are interested in green stuff I try and get them to join the committee or join the Facebook page.

Talia also uses social media to both influence and educate by linking the broader issues with the day-to-day things that happen in the lives of her friends. Talia said:

Showing them how it impacts their life, or might impact their life, then they start to trigger that empathetic response. With me, a lot of girls my age are in to fast fashion so I can tell them about the problem and then share links to cool ethical companies and take friends out op shopping [for second hand clothing] for the day.

Interviewees spoke of creative ways of sharing information such as practical demonstrations, graphs or visuals, photos, social media, musical representations, and other ways of communicating a vision of the future. For Thiha, forced to be physically separate from allies, social media has to take the place of talking, but he still finds emotional hooks that tie humans together socially as a way of promoting awareness and potential action. Thiha said:

We share that experience and information secretly in Burma and now openly at conferences and during refugee week. We are promoting refugee stories, like my own story, to explain to people what is going on. We have media involved, storytelling sessions, campaigning, publishing articles to spread the information. We publish information on websites and blogs, online platforms, LinkedIn, Facebook, Twitter [X].

Enact

Allowing the work to speak for itself was a common suggestion from the interviewees. Many purported that a positive attitude is infectious and that demonstrating practical examples was a valuable strategy. They said that they often did not need to say anything specific about what they were doing but allow their actions to do the talking for them. Mehreen shared how she engages her family members:

With my sister, she would see me doing what I was doing and I would just take her along... so she learned by seeing, then a little bit by doing. With my partner, I said "I'm going to do this... No force or pressure, just we're looking for this and you have those skills, are you interested?" Then if they do, great, if not, that's okay too... No force, gentle. I keep doing what feels right to me. I tell people what I'm doing, sometimes they show up and sometimes they don't. If it makes sense, people will join. Doing what I do, talking about it, and sharing.

Ben tries to be a good example to other people through how he lives his life; his behaviours are reflective of his values. He thinks about what motivates him personally and applies those same encouraging strategies when engaging with other people. He finds he is most encouraged when he can see that there are solutions to problems. He tries to understand the details and when he talks with others, "I try to give good information, I try and be positive, I try and be solutions-based, give resources." Vijeta identifies as an introvert, but she takes inspiration from Gandhi, "You can shake the world in your own quiet way. You don't have to be loud... So just be you quietly." Aruna similarly said, "Be the change you want to see... I really, really love that and that is something I hold with me every time when I work. I show them the impact. And the impact is visible."

As a high school teacher Jenny takes informal opportunities to encourage people to join her everyday actions:

> Yesterday I went to class in my new op shop outfit. Before we started the class, I had a chat to them about op shopping and how it supports a good organisation, gets away from slavery, and reduces contributions to landfill. It's good to be that model of doing it and presenting it in a fun way. It is good to use teaching as a platform. The kids are so open and willing to take on new ideas.

Caroline made this recommendation:

> You have to prove to them that you're doing things before being able to get more people to join... To get people to participate you have to show it is working, that you're getting traction... I've learned to start with some small wins... don't go straight away to a big project. And thanks to small projects, they will bring small victories, but all together will give a good picture. This can prepare you so you can see what you need for a bigger project.

Nicole had a similar strategy but works at different pace, "I'm on a train and it's going full speed ahead to destination greatness. Everyone knows what my aim is – I'm just like – 'Jump on. Jump on.'"

Finding ways of inviting and engaging others is a core component of how the world is going to address our collective issues. Journalist Malcolm Gladwell (2013), in his book *The tipping point – How little things can make a big difference*, draws upon a biological analogy, an epidemic, to understand how ideas, products, messages, and behaviours can spread like viruses. The underlying characteristic is contagiousness – little changes can have large effects resulting in a dramatic change when a *tipping point* is reached (Gladwell, 2013, p. 9). Author Ruha Benjamin (2022) in her book *Viral justice: How we grow the world we want*, also draws upon the analogy of the virus. Learning from how COVID-19 spread through-out our communities, she asks the reader to consider this to be a model for spreading justice. Her vision involves starting with our own families and communities, accepting our interconnectedness, doing a little at a time, contributing daily, identifying the underlying causes, and then enacting change where we can. She writes,

> A microscopic virus has news for us: a microvision of justice and generosity, love, and solidarity can have exponential effects... We can be one of them, if we choose: vectors of justice, spreaders of joy, transforming our world so that everyone has the chance to thrive.
>
> (Benjamin, 2022, p. 16)

Personalise

Connecting with people on an emotional level is a key element of a suc-cessful recruitment strategy. A personal one-on-one approach is considered from a political, marketing, and management perspective to be the best way of encouraging potential activists to become involved (Pettitt, 2020). This can be followed up with invitations to social events because friend-ships are one of the best ways of retaining people (Bowman, 2019). Psy-chologists Miklikowska and Tilton-Weaver (2022) support this with their research on intergroup relations, anti-immigrant attitudes, diversity, empa-thy, activism, and adolescent development. Research shows that peers are more likely to engage in behaviours such as volunteering when their friends are volunteering (van Goethem et al., 2014).

However, engaging one-on-one can be an uncomfortable prospect if these types of interactions do not come naturally. Fortunately, research indicates that emotional intelligence can be cultivated. Researcher Dr Marc Brackett (2019) has shown that emotional intelligence is malle-able and has identified the factors that influence the degree to which we can express our emotions in authentic and honest ways. Professor of psychology Jamil Zaki (2020) draws on tools from psychology and neuroscience to argue in his book, *The war for kindness: Building empathy in a fractured world*, that empathy should not be considered a genetically hardwired trait. Rather, it is possible to learn how to empathise more effectively.

Interviewees said they had to find what motivated the person they were dealing with and make an individual connection. It was not a standar-dised approach but a refined nuanced way of interacting with others to reach them personally. Interviewees had to connect to that person's life experiences, use their people reading skills, understand their drivers, and then find a hook that reaches that person. Interviewees emphasised that the strategies were careful, gentle, incremental, diplomatic, and quiet. They suggested finding a topic that the person was connected to or interested in and then support them to see how they could contribute in a tangible way. This involved making the activities and process meaningful, as well as helping them identify personal benefits.

For many interviewees, sharing information to connect with people on an emotional level predominantly meant talking. Fostering calm dialogue, exchanging points of view, taking time for meaningful conversations, telling stories, and speaking from the heart were examples. Clariana said that people's mindsets can be influenced by explaining information in relatable ways and infusing conversations with an appropriate level of passion: "I love talking about it and being like 'Have you thought about this?' I think information is such an underrated aspect of advocacy. When people learn and they are willing to learn, that's when you change people's mindsets." Iram focuses on sharing personal stories with her friends and family. She said:

> I share my work and experiences with friends and family to [share] how the world is different for different groups and communities. Many of my friends work in corporate sectors and many at times are not aware of many existing social issues and difficulties faced by certain groups. My sharing helps them to understand it better. They sometimes take my stories and use them as examples in their platforms.

Alicia finds that personalising her work is a good way to engage people in topics they may not necessarily want to discuss. To tap into the emotions of people in leading or influential roles in organisations, she incorporates her personal story and those of women with whom she had worked: "I needed to share Ellie's story to show the gravity of the risk that the frontliners take." Additionally, Serena said, "Storytelling from your own personal experiences is the best form of mentoring and bringing others along that journey." Ekawati also shares stories with people as a means of connecting them emotionally to the project: "Small stories help. I tend to share stories before asking people to join, help to form pictures they can connect on their own later."

Ricki suggested that it is important to take the time to understand who she works with and what they want to achieve: "You've got to understand your audience... you have to find out what floats their boat... Talk in the language they're happy with.... You have to understand where the person is at." Susanne felt it was important to get to know people to find out how to motivate them: "Like with the CEO, getting him to understand how this work will help position us well in the sector [involved showing him] the type of evidence that I think will motivate him." Aruna said that although she quotes statistics about the level of need and talks about how change is possible, she always personalises the interaction by bringing the audience into the story. When Adrian interacts with other grandparents, he encourages them not to think about themselves, but suggests they focus on generating hope for the next generation. He finds the hook that helps co-workers think beyond their own issues and focus on something they are emotionally connected to in the future. Adrian said, "I think every individual person has got their own hot buttons, and cold buttons as well. The on switch as well as the off switch, and it's not always the same switch."

An example of the potential impact for such actions comes from Natalie Isaacs (2022) and her book *Right here right now: How women can lead the way in the climate emergency.* She shares how she became motivated to make personal changes to address climate change. Previously disengaged and feeling disempowered, after making small changes in her household, she found the confidence to interact with other disengaged women and encourage them to empower themselves to change their lives and habits. Isaacs left her glamorous career in the cosmetics industry to start an organisation called *1 million women* (Isaacs, 2023). The organisation, now with over one million women committing to raising awareness, is predicated on the belief that conversations among people who know each other is a catalyst for mobilisation and individual agency (Isaacs, 2022, p. 82).

Dynamics

Research shows that there are five key elements that consistently appear in groups that work cooperatively (Johnson & Johnson, 2014, 2015). When these five elements are purposefully combined, as in a recipe, successful social relationships are more likely to achieve the cooperative goals. These five elements provide a useful framework for thinking about the *ingredients* required to support positive group dynamics. According to academics Johnson and Johnson (2014, 2015), the five elements of cooperative teamwork include:

1   Establish a shared goal where team members rely on each other to achieve success.
2   Be accountable to each other by explicitly allocating roles and doing one's part.
3   Provide encouragement through helpful behaviours and offering support.

4   Use social skills to communicate inclusively, manage conflict, and build trust.
5   Reflect on what is working well for the group and what needs adjusting.

Establish goals

This ingredient is about establishing a situation where everyone in the team believes that they could not achieve their ambitions without their co-workers also succeeding. The team needs to be oriented toward the same shared goal or end state. Resources, roles, tasks, and rewards are linked, and team members understand that what they are contributing is integral to achieving the desired success. There needs to be alignment between what the individual wants, what the individual can contribute, and how the group defines success. When the overarching vision aligns between group members, and they see how their work is essential to achieve the win, then great things can happen. Johnson and Johnson (1989, p. 75) state, "When group members recognise that they 'sink or swim together', feelings of personal responsibility to do one's fair share of the work and to promote the success of fellow group members results."

One of the main ways interviewees motivated others was by sharing the overarching vision and common values in a way that connected with the individual with whom they were engaging. Focusing on the goal, helping them find where they could contribute, and facilitating opportunities for people to identify synchronicity with others was important to motivate their team. This was articulated by Susanne who said, "[To] help people see how they can influence and make change, I try to paint for people a vision about how things can be... try to create a vision or working with people to create a vision."

Understanding the underlying motivations and identifying what motivates their group members meant interviewees could tap into their passions and let those be the reason for participation. Whether they were trying to bring people along with them or influence people in positions of power, developing meaningful interpersonal relationships was the key to solidifying mutual goals. Ronny articulated his approach to finding the shared goal:

> Your struggle is my struggle – I always approach it in a way that our struggles are bound in each other or intertwined. So that means that when anyone joined the cause, they know that it's also their struggle, and it's something they're passionately advocating for, whether it's the land rights or it's sovereignty or it's climate change. And so, that's where we find that common ground, and we share that together. And then how we build the capacity of that,

the movement or the campaign together – in that way, we are in this together.

Liz's tree planting activity demonstrates how she shared her vision with others and modelled the change she wanted to see to motivate others. She said:

> When you're doing this kind of activism work, the law of attraction actually works. You will always find some people who have your vision, it may be one or two people but you'll find the vision... I feel like the law of attraction has actually worked a lot in my vision.

Even when there was no interpersonal connection or opportunity for face-to-face engagement, some still shared how important it was to ensure people felt part of something bigger. Alyssa finds that she can help people develop a shared vision and establish a mutual goal through social media. She said, "Social media is really good at reaching out with people. I have a lot of communications over social media. Trying to bring people on board and make them feel like they're part of something."

Encourage

This ingredient in the recipe for team success is about group members encouraging and supporting each other to achieve. Examples include assisting, helping, exchanging ideas and resources, providing feedback, discussing, debating, advocating on each other's behalf, guiding, influencing, decreasing anxiety, and trusting each other. It is about making sure members of the group know each other well and understand where each other is coming from on a personal level (Johnson & Johnson, 1989). Trust is the fundamental concept because to share thoughts the group members need to believe that they will be treated with respect; team members need to have confidence in their peers that they will behave in ways that will benefit each other.

Interviewees reported that creating a positive atmosphere where people were encouraged to have fun, enjoy themselves, and celebrate their achievements, meant people were more likely to join. Kim thought that "enthusiasm, passion, kindness" were important as well as "valuing their input and showing that you value it". Tamsin also believes that offering encouragement is a good tactic as she finds that people respond well when she tells them how fantastic their contribution is no matter how small. Marilee had some excellent strategies for providing the right amount of encouragement to motivate others. She suggested a very targeted and nuanced approach:

> I like to listen to people when they're talking about what they're passionate about and I think that encourages more, it's good positive reinforcement. Whenever someone shows an interest in something, even if it's not something I'm interested in, I try and back them up and promote them and cheer them... I take a pretty wide view of what a contribution looks like, and I'm really accepting – if people do a tiny little thing, I'm excited for them and I cheer them on. And the same if they're doing a huge thing, like a career change, I cheer them on then too just as excitedly. I demonstrate leadership and excitement about making changes.

Alex sent out monthly emails at her previous workplace with updates on the amount of waste that had been diverted from landfill along with positive messages about the teams' efforts and impact. She was always positive and never included a negative tone. Alex said that this "made me the waste queen and made people feel like I was approachable to discuss waste and recycling".

Khalil noted the importance of providing encouragement with the purpose of highlighting the mutual wins and shared gains:

> People have to see the reward of their work. If people see that the movement is being successful, then people love to join. When people see something beautiful and shaping up, then this is a magnet. People don't want to join a sinking ship, they want to join a fast aeroplane. So you have to lead by example, do things successfully, invite people.

This is particularly important for long term projects. Elise highlighted the importance of providing the right amount of encouragement along the often drawn-out slow burn campaigns for change:

> When you're working on these longer-term campaigns... the struggle for charities or groups who are trying to maintain a more ongoing level of engagement and activism is how you create those moments that give people that sense of a win and being able to kind of pull back rather than it being this kind of constant grind.

Janelle provided excellent advice about changing the tempo and offering variety to manage this issue of waning support over the long term.

She gave an example of how to foster a positive collegial working environment:

> We spent a lot of time building trust with each other and not necessarily always doing the hard activist work but doing things like having a stall at a market and making a banner and going to a protest and stuff and then go to the bowls club for a beer afterwards. We put a lot of effort into that [building social connection and trust] and it paid off.

Maintaining momentum and providing encouragement in the form of gentle tenacity and perseverance was noted by Daisy:

> I encourage individuals, there is no amount too big or small that you can do. If you have a vision, if you like to do something, you just simply need to help one another and just simply have the courage to do it... so if we have challenges, if we have successes – we share it – it doesn't matter how big or small what they are – to just keep the spirit of the staff to keep moving.

Be inclusive

If group members lack social skills then interactions are unlikely to result in effective teamwork (Johnson & Johnson, 1989). Social skills can include interpersonal and small group skills such as communication, leadership, decision making, trust building, distributing power and influence, and conflict management skills. Co-workers must have or learn these skills and be motivated to use them. If there is a high level of social skills across the group, then less skilled members will learn from those with more developed skills (Johnson & Johnson, 1989).

Working in cross-cultural situations and with different types of people requires specific social skills. Preferred learning styles, ways of thinking, and ways of communicating will differ when working with people who are neurodivergent, from different socioeconomic backgrounds, have different levels of education and aptitudes, have varying levels of ability, and are of different genders and ages. It is about recognising the urgent need for everyone to consider how they can promote diverse and marginalised voices. Thinking about how to work with more humility and wisdom to support interactions among diverse team members is important. Community and international development specialist Deborah Rhodes (2022) suggests that people need to reflect on their roles, values, and levels of power when working in cross-cultural or multi-cultural contexts. Celebrating cultural expression and valuing the input of diverse ideas requires activists to recognise that our beliefs about the way change happens can vary across cultures. Many different resources exist to help us develop skills to be more inclusive. Some resources provide highly nuanced suggestions about how to engage with others, while others provide practical suggestions for how we can tweak the way we operate to make interactions more welcoming and effective (Blue Sky Community Services, 2016; Heath & Wensil, 2019; Inclusion Solutions, 2023; Inclusive Australia, 2021).

Interviewees said that ensuring an inclusive, safe space for participation was essential. Accepting differences and encouraging the incorporation of diverse opinions were strategies for engaging people on an ongoing basis. Kim suggested that "you get so many important ideas and suggestions by empowering others. Empowering others builds a much stronger, more inclusive, more effective team." Jody's example exemplifies this sentiment:

> I'll bring them to the table and sit and listen and hear… I'll bring them along the journey and ask for their input. Because I mean at the end of the day it's providing a seat at that table… Providing a platform where their voice is heard or their ideas, and that they're valued. I think the most important thing to me is listening because when I listen, I think I learn more.

Jody also understands the importance of making sure people from all different cultural backgrounds are invited to have a seat at the table. She said that no matter what your own background, everyone will need to engage in cross-cultural situations. Jody suggested that recognising our own biases was an important first step:

> We've all got it, we've all got bias, implicit bias, unconscious bias. Or conscious, for that matter. Education is the way to go to addressing. We all can learn and understand each other's culture. I know when I used to travel internationally you want to be respectful to the country that you're going to. So, this is our Country and we're here to share it with everybody.

Offering opportunities to people from different backgrounds and circumstances was also important. Aruna's example of how she engages with a diverse group of young people provides a practical example. To ensure their involvement in her work, she includes them in every step of the process, "The logos, the wordings, the promotions – because I know I can bring my experience in – but I can't bring the youth language into my work." Similarly, Ronny takes a very interactive approach to involve people in his campaigns. He thinks about it as building the capacity of everyone involved – including himself. He said:

> In that way, we are in this together. It's not that, "Oh, can you help us?" that approach of helping us – begging… In the music space, it's like, "Okay, let's write music and let's perform together." So artists… they use their skill set. They use their knowledge in that space, which is shared benefits. We share that together. So, we all feel that we are part of that movement together.

Be accountable

Being accountable and holding other people to account is one of the trickiest ingredients to incorporate into the recipe. This is especially the case for everyday activists who may not be in positions of power or who are working in organisations with flat organisational structures. This element is about making sure that each person on the team knows that they have a responsibility to complete the task they have been allocated (Johnson & Johnson, 1989).

A person in a position of power may be able to employ someone to do a specific task, change a job description, or add actions to a structured work plan. An everyday activist working with a group of colleagues may not have access to any of these mechanisms. Yet it is still important to find practical ways of ensuring that co-workers take responsibility for certain tasks and negotiate ways to hold themselves accountable to these commitments. If people leave meetings without understanding what, when, and who is responsible for what tasks then the members of the group will not feel that tangible sense of interdependence which is important for fostering positive group dynamics. Unfortunately, none of the interviewees provided concrete examples of strategies that align with this element, but this is understandable when we think about how difficult it is for everyday activists to hold their colleagues to account.

Historically, there has been a lack of research focused on people who are not in leadership positions, which has resulted in limited knowledge about the potential of these people to influence group dynamics (Uhl-Bien et al., 2014). However, that situation is being addressed with increased focus on people who are not aspiring to be in official leadership positions, but who want strategies for working with others in creative, innovative, and productive ways (Madden, 2011; Riggio & Chaleff, 2008). These strategies highlight that mechanisms of accountability are not necessarily related to compliance, pressuring people, or applying authority. Rather, they are about planning together to ensure clarity around purpose and expectations, that people have the capability, willingness, and capacity to achieve what is required, and that the environment supports people to take ownership for their own accountability (Rockwell, 2015).

Reflect

The element of reflecting on how the group is functioning is another ingredient that is often overlooked (Johnson & Johnson, 1989). Considering what actions were helpful or unhelpful, what needs to continue, and what should be changed are key questions. This element can help clarify the purpose and roles of group members and make improvements regarding the effectiveness of their contribution. A continuous improvement process can help to keep things on track, avoid the dynamics from derailing into negativity, and find ways that the group work can be enhanced. Reflecting regularly on the process can ensure the skills available are used in the most effective way possible. Johnson and Johnson (1989) state that:

> It is a truism in group dynamics that to be productive groups have to "process" how well they are working and take action to resolve any difficulties members have in collaborating together productively.
>
> (p. 75)

Interviewees suggested avoiding trying to control everything; having infeasibly high standards is a sure way to turn people away. Kim advised against being autocratic while Nigel offered this piece of advice:

> Try and give people more space and not be as much of a control freak... That's a major problem for a lot of advocacy organisations. You get one or two very strong-willed, strong-minded people who basically say, "It's my way or the highway" and people will join and fall away because they can't find space – they're not prepared to just go along with that dominant person's way of doing things.

The interviewees recommended taking the time to ask group members what they find satisfying about their roles, what could be improved, and what they require to develop their skills. Research shows that this type of information can not only help prevent team members from becoming dissatisfied but provide a rich source of potentially untapped resources that can support the development of the group. Identifying the struggles, seeking ideas for success, finding the right balance of support and autonomy, and determining the factors that will lead to increased levels of satisfaction are some of the benefits of group processing (Prince & Piatak, 2022).

In summary, establishing shared goals, providing encouragement, using social skills to be inclusive, finding ways of holding people accountable, and taking time to reflect on how the group is functioning are the five key elements required for positive group dynamics. However, despite all good intentions, working with humans means that tension and disagreements will occur, so in the next section we provide some strategies for dealing with these situations.

Manage conflict

People working on a cause will most likely have different ways of understanding how change happens, use different methods, and apply different strategies. The nature of the work involves bringing people together who have diverse worldviews, varying value systems, and who work using a myriad of approaches. Romano et al. (2021) state,

> Activist networks are complex and dynamic, with variable communication pathways, a high degree of turnover of members, varying degrees of physical distance between actors, and at times a lack of transparent organizational structure within those networks; all of which can generate conflict.
>
> (p. 298)

The early signs of conflict such as forcing, withdrawing, avoiding, and making demands need to be monitored with vigilance to prevent escalation. Try to deal with conflict immediately and attempt to find solutions when the first signs appear because preventing issues from escalating, navigating through conflict, and dealing with the repercussions of negative events can take up time, drain energy, divert resources, and detract from achieving the goal. These emotionally charged situations, that may have been initiated by seemingly insignificant differences of opinion, can rapidly turn into worst case scenarios.

Finding creative ways of making decisions collectively can be a useful way of avoiding conflict. Reaching consensus contributes to ensuring group members are in control and that the power is shared more equitably (Ciccarone et al., 2022; Seeds for Change, 2023). Conflict is not always negative. Differences of opinion could be considered normal and part of a robust way of working that does not have to relate to the underlying personal relationships (Fujimoto et al., 2014).

Beneficial opportunities can arise if debate is seen as a healthy way of incorporating diverging viewpoints (Hyde, 2012). As Ricki stated:

> I don't want everyone to agree with me. I like diversity. It's about respecting diversity, having a different opinion, working in harmony, not silencing the difference... It's about empowering and building resilience... You have to have people in different spaces doing different things to achieve change. You have to have the visionaries and you have to have the micro where they're working on the daily interactions and getting the dialogue happening.

Another strategy includes pre-planning as a group around how to deal expediently and constructively with conflict. Clarifying and examining the underlying needs, feelings, fears, and anxieties of those involved can give all group members the information required for harmonious collaboration (Tjosvold et al., 2014).

Interviewees provided some warnings on the topic of managing conflict. They said that resistors should not be allowed to infiltrate the group dynamic or threaten the potential for goal achievement. Interviewees recommended avoiding people who may sabotage the work because these are more dangerous than people who simply resist. The interviewees were prepared to politely ignore resistors but they could not stand by when their work was being actively threatened. Khalil made this recommendation:

> There are people who will be motivated by themselves and for their own interests. If you see people coming only for their own self-interest, as a leader you need to manage those people, to protect the project, the idea, from being exploited or sabotaged. I think we don't give attention to that sometimes. Everyone thinks that when you volunteer there's no vetting, no skills needed, and that's not right. We need to be a bit selective. Sometimes people are not coming to help the cause, they're helping themselves.

Conflict resolution skills can be learnt, just like other social skills such as communication, facilitation, and cultural competence (Johnson & Johnson, 2003). Monitoring personal reactions is also an important factor. Knowing yourself and your own inclinations, work preferences and triggers, as well as understanding these characteristics of your team members, will affect your ability to get the most out of a team. It is not only about others, but also yourself. Reacting negatively when conflict arises could be an opportunity for growth and improvement.

You may have to pick your battles, be willing to forgive, and let some things go that are not important. Being open to changing your opinion or finding alternative ways of achieving goals can be ways of demonstrating flexibility. Negotiation and facilitation of mutual problem solving are skills that need practice and honing. The first proposed solution may not be the most effective, so trial strategies and seek information to see whether the change is what was expected. Khalil suggested that a degree of self-reflection is required to ensure that leadership of the group is rotated to share power in a distributive way:

> But you shouldn't stay in the thing for a million years, you become like a dictator, you need to pass to the next person. The best thing you can do is to go away when it is good timing to go. Power corrupts and they want more things. I think about activism in a collective way, within networks, working with other people. I think some very important elements of democracy and rotating leadership are useful, not only for motivating people but also for making things work. Otherwise, if you just keep holding on, it's very fatiguing and it gets very boring, for you and for other people. When you bring new people they come with fresh ideas and it can be very beautiful again.

Whatever mechanisms are selected for conflict resolution, these will be an integral part of strengthening communication and cooperation, transforming the broader societal systems addressing the underlying power dynamics, connecting with allies, and understanding differences among groups beyond the immediate team (Romano et al., 2021). Seeking out online resources can be a quick way of refreshing strategies and finding inspiring ways of considering challenges (Community Door, 2023; Community Industry Group, 2023; New Zealand Government, 2023).

Reflection

Bringing people we know along on the journey is an excellent way of garnering support and increasing the amount of work we can achieve. Activists do not have to just hope or wish for a great team; thinking intentionally and strategically about collaboration can decrease stress and enhance the likelihood of cooperative, effective, and uplifting group dynamics. The underlying message from the interviewees, which is supported by research, is that meaningful connections with the people around us are the most effective way of expanding our network. Based on social psychology, the *recipe* described in this chapter provides a useful framework for everyday activists to plan their interactions with co-workers (Rogers et al., 2021). These questions placed against each element of the recipe are prompts for consideration.

Establishing shared goals

- Have you clearly articulated your goal?
- Do team members agree on what they are working towards?
- Does the team have long and short-term goals and descriptions of what success would look like?
- Have you collaboratively developed a plan that connects individual roles to the goals?

Providing encouragement

- How does the group provide recognition to individual members?
- Does the team trust each other enough to provide honest feedback?
- Are there milestones that are celebrated along the way?
- Does the team have opportunities to engage socially?

Developing social skills

- Are different styles of communicating, learning, and engaging incorporated?
- How do you welcome and encourage diversity (including cultural, neurodivergent, (dis)ability, sexual, gender, socioeconomic, age, etc.)?
- Is there an agreed way of dealing with conflict when it arises?
- Is information communicated clearly and tailored for group members?

Being accountable

- What mechanisms do you use to hold yourself and others accountable?
- How does the group follow up on task progress?
- Does the group have a system to manage information?
- Do agendas, minutes, or communication logs make decisions transparent and accessible?

Reflecting on the process

- Do you take time to consider whether the group is functioning well?
- Do you have a process for finding out from group members what they think?
- Are there scheduled opportunities for reflection?
- Do decision making processes include a feedback mechanism?

Conclusion

Cooperation is the platform upon which everyday activists need to base their initiatives if they are to be effective. Activists need to harness interpersonal interactions to work with others to accomplish goals and ensure that everyone wins. Although survival of the fittest and a competitive view has dominated thinking for the past century or more, a renaissance of work across academia and popular culture has subsequently focused on interdependence (Kropotkin, 2012/1902; Servigne & Chapelle, 2022). Professor of animal behaviour, Dr Ashley Ward (2022a) argues that we should not think of ourselves as superior to the rest of the animal kingdom but instead take lessons from them about cooperation (Ward, 2022b). Expanding the concept of connection among humans, Benjamin (2022) reinforces our interdependence with everyone and everything. She writes that to oppose ableism, homophobia, transphobia, racism, sexism, classism, and the effects of colonialism, we are required to:

> Acknowledge and foster a *deep-rooted interdependence*, not as some cheery platitude but as a guiding ethos for regenerating life on this planet. This is what disability justice organizers have been trying to tell us, and what Indigenous peoples have long asserted – that whether we want to accept it or not, we are connected, not just to living things but to those yet born… Interdependence is not only part of a sacred philosophy but also a guiding ethos for refashioning social and political structures.
>
> (Benjamin, 2022, p. 10)

References

Benjamin, R. (2022). *Viral justice: How we grow the world we want.* Princeton University Press.

Blue Sky Community Services. (2016). Social inclusion and diversity. *Blue Sky Community Services.* https://bluesky.org.au/communities/social-inclusion-p roject-where-everyone-belongs/.

Bowman, B. (2019). Imagining future worlds alongside young climate activists: A new framework for research. *Fennia - International Journal of Geography*, 197(2), 295–305. doi: https://doi.org/10.11143/fennia.85151

Brackett, M. (2019). Become an emotion scientist with Marc Brackett. *The Psychology Podcast.* https://scottbarrykaufman.com/podcast/become-an-em otion-scientist-with-marc-brackett/.

Ciccarone, M., Nayak, P., Araia, Y., & Seeman, B. (2022). Five ways non-profits can make decision making more inclusive - and more effective. *The Bridgespan Group.* https://www.bridgespan.org/insights/5-ways-nonprofits-ma ke-decision-making-inclusive.

Climate for Change. (2023). *Climate for Change.* https://www.climateforchange. org.au/socialresearch.

Community Door. (2023). Resolving conflict. *Community Door.* https://comm unitydoor.org.au/resources/human-resource-management/resolving-conflict.

Community Industry Group. (2023). Top tips for managing and resolving conflict. *Community Industry Group.* https://communityindustrygroup.org.au/ topic/top-tips-for-managing-and-resolving-conflict/.

Deutsch, M. (1949). An experimental study of the effects of cooperation and competition upon group process. *Human Relations*, 2(3), 199–231.

Deutsch, M. (2011). Cooperation and competition. In P. T. Coleman (Ed.), *Conflict, interdependence, and justice: The intellectual legacy of Morton Deutsch* (pp. 23–40). Springer. doi: https://doi.org/10.1007/978-1-4419-9994-8_2

Effective Activist. (2023). *Effective Activist.* https://effectiveactivist.com/movem ents/recruitment/#_ftn8

Espen Stoknes, P. (2015). *What we think about when we try not to think about global warming: Toward a new psychology of climate action.* Chelsea Green Publishing.

Franco, M. (2020). How to make friends as an adult. *Psyche.* https://psyche.co/ guides/how-to-make-new-friends-when-youre-busy-with-adulthood.

Fujimoto, Y., Rentschler, R., Le, H., Edwards, D., & Härtel, C. E. J. (2014). Lessons learned from community organizations: Inclusion of people with disabilities and others. *British Journal of Management*, 25(3), 518–537. doi: https://doi.org/10.1111/1467-8551.12034

Gladwell, M. (2013). *The tipping point: How little things can make a big difference.* Abacus.

Green, D., & Gerber, A. (2015). *Get out the vote: How to increase voter turnout.* Brookings Institution.

Heath, K., & Wensil, B. (2019). To build an inclusive culture, start with inclusive meetings. *Harvard Business Review.* https://hbr.org/2019/09/to-build-a n-inclusive-culture-start-with-inclusive-meetings

Hronis, A. (2022). Why do we find making new friends so hard as adults? *The Conversation.* https://theconversation.com/why-do-we-find-making-new-friends-so-hard-as-adults-171740

Hyde, C. A. (2012). Ethical dilemmas in human service management: Identifying and resolving the challenges. *Ethics and Social Welfare*, 6(4), 351–367. doi: https://doi.org/10.1080/17496535.2011.615753

Inclusion Solutions. (2023). Empowering inclusive communities. *Inclusion Solutions*. https://inclusionsolutions.org.au/.

Inclusive Australia. (2021). Inclusive Australia. https://www.inclusiveaustralia.com.au/about-us

Isaacs, N. (2022). *Right here right now: How women can lead the way in the climate emergency.* HarperCollins.

Isaacs, N. (2023). *1 million women.* https://www.1millionwomen.com.au/

Johnson, D. W., & Johnson, R. (1989). *Cooperation and competition: Theory and research.* Interaction Book Company. http://psycnet.apa.org/psycinfo/1989-98552-000

Johnson, D. W., & Johnson, R. (2003). Training for cooperative group work. In M. A. West, D. Tjosvold, & K. G. Smith (Eds.), *International handbook of organizational teamwork and cooperative working* (pp. 167–184). John Wiley & Sons. doi: https://doi.org/10.1002/9780470696712.ch9

Johnson, D. W., & Johnson, F. P. (2014). *Joining together: Group theory and group skills* (11th edn.). Pearson Education Limited.

Johnson, D. W., & Johnson, R. (2015). Cooperation and competition. *International Encyclopedia of the Social & Behavioral Sciences*, 2(4), 856–861. doi: https://doi.org/10.1016/B978-0-08-097086-8.24051-8

Kropotkin, P. (2012). *Mutual aid: A factor of evolution.* Dover Publications (Original published in 1902).

Madden, K. (2011). *The synergetic follower: Changing our world without being the leader.* Createspace Independent Publishing Platform.

Miklikowska, M., & Tilton-Weaver, L. (2022). Empathic friends can provide the right kind of peer pressure. *Psyche.* https://psyche.co/ideas/empathic-friends-can-provide-the-right-kind-of-peer-pressure

Minkler, M., & Wakimoto, P. (Eds.). (2021). *Community organizing and community building for health and social equity.* Rutgers University Press.

New Zealand Government. (2023). *Manage conflict in a group.* Department of Conservation Te Papa Atawhai. https://www.doc.govt.nz/get-involved/run-a-project/community-project-guidelines/manage-conflict-in-a-group/

Pettitt, R. T. (2020). Recruiting and retaining party activists. In *Recruiting and retaining party activists* (pp. 1–20). Springer International Publishing. doi: https://doi.org/10.1007/978-3-030-47842-1_1

Prince, W., & Piatak, J. (2022). By the volunteer, for the volunteer: Volunteer perspectives of management across levels of satisfaction. *Nonprofit and Voluntary Sector Quarterly*, 089976402211279. doi: https://doi.org/10.1177/08997640221127974

Rhodes, D. (2022). *Facilitating change across cultures: Lessons from international development.* Practical Action Publishing.

Riggio, R., & Chaleff, I. (Eds.). (2008). *The art of followership: How great followers create great leaders and organizations.* Jossey-Bass.

Rockwell, D. (2015). *How to hold people accountable without using authority.* Leadership Freak.

Rogers, A., Kelly, L., & McCoy, A. (2021). Using social psychology to constructively involve colleagues in internal evaluation. *American Journal of Evaluation*, 42(4), 541–558. doi: https://doi.org/10.1177/1098214020959465

Romano, A., Linnemeier, E. K., & Allen, S. H. (2021). Conflict resolution in activist networks: Reflections on prework, power, and engaging with change in Appalachia. *Conflict Resolution Quarterly*, 38(4), 283–302. doi: https://doi.org/10.1002/crq.21303

Seeds For Change. (2023). *Consensus decision making.* Seeds For Change: Strengthening Co-Operation, Empowering Resistance. https://www.seedsforchange.org.uk/consensus#stages

Servigne, P., & Chapelle, G. (2022). *Mutual aid: The other law of the jungle.* John Wiley & Sons.

Shenker-Osorio. (2021). *ASO Communications.* https://www.asocommunications.com/

Tjosvold, D., Wong, A. S. H., & Feng Chen, N. Y. (2014). Constructively managing conflicts in organizations. *Annual Review of Organizational Psychology and Organizational Behavior*, 1(1), 545–568. doi: https://doi.org/10.1146/annurev-orgpsych-031413-091306

Torca, A., Mazei, J., & Hüffmeier, J. (2022). Why some teams boost motivation while others totally sap it. *Psyche.* https://psyche.co/ideas/why-some-teams-boost-motivation-while-others-totally-sap-it

Uhl-Bien, M., Riggio, R. E., Lowe, K. B., & Carsten, M. K. (2014). Followership theory: A review and research agenda. *The Leadership Quarterly*, 25 (1), 83–104. doi: https://doi.org/10.1016/j.leaqua.2013.11.007

van Goethem, A. A. J., van Hoof, A., van Aken, M. A. G., Orobio de Castro, B., & Raaijmakers, Q. A. W. (2014). Socialising adolescent volunteering: How important are parents and friends? Age dependent effects of parents and friends on adolescents' volunteering behaviours. *Journal of Applied Developmental Psychology*, 35(2), 94–101. doi: https://doi.org/10.1016/j.appdev.2013.12.003

Ward, A. (2022a). *The social lives of animals: How co-operation conquered the natural world.* Allen and Unwin.

Ward, A. (2022b). What can animals teach us about social cohesion? Radio National, ABC. https://www.abc.net.au/radionational/programs/latenightlive/what-can-animals-teach-us-about-social-cohesion/13797442

Zaki, J. (2020). *The war for kindness: Building empathy in a fractured world.* Broadway Books.

6 Dealing with people who disagree

The amount of energy that is potentially wasted on people who have different points of view can significantly impact the work of activists and be a barrier to achieving ambitions. Making decisions about whether to interact at all, what strategies to use, how to identify the right time to disengage, and how to determine what to do next, does not have to be guesswork. The everyday activists that were interviewed for this book have a range of useful hints about how they handle these issues. Research from the behavioural sciences and philosophy, and literature from popular culture, also provides guidance. This chapter aims to support activists' confidence about knowing when to disengage and when to engage. We acknowledge at the outset that there is a myriad of reasons why people might disagree with our stance. However, in this chapter we take the extreme end of the spectrum focusing on denialism as a case study for the purposes of examining how we as activists can effectively deal with people who disagree with our point of view. We outline the benefits that can result when people who disagree, or who may be in denial, are challenged about their beliefs. The purpose of the chapter is to highlight the importance of listening to understand more about the beliefs of others and identify our own biases. The chapter concludes with suggestions to further develop interpersonal communication skills, courage, and self-reflection.

Understanding and engaging with people in denial

As activists are working to change the current situation, interactions with people who want to retain the status quo are almost inevitable. Regardless of the topic, it is likely that there will be controversial aspects of activists' desired goals or processes, to which sections of society hold strong emotional attachment and about which they have beliefs, value systems, and social networks supportive of their opinion. These individuals could be

DOI: 10.4324/9781003333982-6

people who simply disagree, who are working to strategically prevent the change, hold conspiracy theories that powerful small groups are responsible for an issue, or do not consider the issue important. There are also people who could be in denial. According to the Macquarie Dictionary (2016), to be in denial is to contradict a statement, refuse to believe a doctrine, disbelieve in the existence or reality of a thing, or to refuse to recognise or acknowledge a particular declaration. To be in self-denial is to close one's mind to an unpleasant fact or experience. Denialism or negationism is also defined as a rejection of "the facts of history, often using unsupported or false evidence, and illogical reasoning" (Macquarie Dictionary, 2016, p. 1).

Historically, science was accepted as providing useful evidence to inform decision making, but in more recent decades some people fear science because they mistrust the institutions associated with science and believe that it can be politically motivated (Spector, 2009). An acceleration of "often educated, smart, and influential people" are embracing dangerous ideas even when "there is nothing to substantiate these beliefs, and easily available evidence shows that they are actually false" (Nadler & Shapiro, 2021, p. 1). Sociologist Keith Kahn-Harris (2018), in his book *Denial – The unspeakable truth*, identifies different types of people in denial. Some are sceptical of all knowledge that comes from established institutions, while others resist one specific body of knowledge. Others passively, privately, and secretly consume denialist information. Some proactively develop information that is anti-establishment and loudly proclaim their beliefs. He argues that the thing that all denialists have in common is that they desire something not to be true. Kahn-Harris (2018) writes:

> Denialism offers a dystopian vision of a world unmoored, in which nothing can be taken for granted and no one can be trusted. If you believe that you are being constantly lied to, paradoxically you may be in danger of accepting the untruths of others. Denialism is a mix of corrosive doubt and corrosive credulity.
>
> (p. 7)

Other writings on denialism expose campaigns that intend to instil doubt and discuss the people who are incited by these campaigns. This literature draws attention to the institutional, political, and financial reasons for the intensification of denialism (Badham, 2021; Farrell et al., 2019; Oreskes, 2019). Developmental biologist Sean B. Carroll (2020) wrote an article in the *Scientific American* that outlines the pattern of predictable denial that underlies most denialist arguments regardless of their topic. He shares the six "principal plays in the denialist playbook" as:

1 Doubt the science.
2 Question scientists' motives and integrity.
3 Magnify disagreements among scientists and cite gadflies as authorities.
4 Exaggerate potential harm.
5 Appeal to personal freedom.
6 Reject whatever would repudiate a key philosophy. (Carroll, 2020, p. 1)

The last line of defence, "Reject whatever would repudiate a key philosophy" can often be the most powerful because it can arrive when all the science is settled. Even when overwhelming evidence exists, denialists can still outright reject the findings because they do not like their implications. If there is no alignment with their worldview or there is a disconnect with their cultural identity, their allegiance to the religious, political, cultural, or social group will prevail over science (Badham, 2021; Carroll, 2020).

However, there is potential for denialists to change their opinions, especially when gently critiqued by people within their social circle. Although trying to debunk the theories of people in denial with more science or literature can result in endlessly spiralling arguments, a critical first step may be recognising, understanding, and anticipating the deployment of the strategies listed above (Carroll, 2020). Another approach is to recognise that denialism can be a reaction to feelings of vulnerability and loss of control in the face of fearful facts (Badham, 2021). Kahn-Harris (2018) suggests that it is worthwhile exploring whether these individuals are actually longing for future chaos, indifferent, secretly frightened by the future, scared that they will lose power, or do not want anything to change.

This chapter shares strategies that the people interviewed for the book and researchers have found useful. It became evident from the responses of the interviewees that effectiveness depends upon the type of denialist; their motivations, patterns, and approaches; and the relationship between the denialist and the anti-denialist. The following sections of the chapter cover when to disengage, listen, connect on a personal level, and when it is appropriate to challenge. We discuss how information can be gained from listening, reflecting, and engaging with people who disagree or who are in denial for the purposes of critically reviewing, planning, and improving. Instead of simply assuming interactions with these individuals are an annoying waste of time, we argue that they can help us break out of our echo chambers and prompt us to research more and think reflexively. If we think deeply about what they are saying, the process may help us better articulate our position and develop arguments with rigour.

Disengage

When we asked the interviewees what tactics they use with resistors and denialists, most suggested not engaging at all; they said not to try, not to waste time. Many interviewees avoid resistors, leave them alone, and believe there is no value in trying to convince sceptics. The consensus was not to confront, or force, or attempt to try to change peoples' minds. Tamsin said, "If they're not interested, I wouldn't waste my time." Iram agrees, "For me, they need to be ignored and I move on." Caroline deliberately chooses to disengage:

> In fact, resistors – I have learned to let them resist. Whatever you know it's like "if you don't want to take care of your planet or your land... I'm sorry, but I'm not going to fight against you". I really only focus my energy on motivating the people who were more like "yeah we can deal with that, who has an idea?" and try to put those in place.

Liz ignores the person and continues on her own path: "When I've had resistance, I don't even try, I kind of just leave them alone and keep doing what I do. I usually don't try to overly convince people, but I keep on the journey." Resonating with this, Janelle said, "I'm a firm believer that [there are those who will] never change, and I can't be bothered dealing with those kinds of people because there's nothing you will ever be able to say or do to change their minds, so it's a waste of time." Mike believes that, although there are some people who are genuinely ignorant and who may change their mind if provided with factual information, most denialists are either not interested in the issues or they are "wilfully determined to have an argument with anyone whose opinion does not accord with theirs." He therefore will only cautiously engage with people who are genuinely open to information through the course of natural conversation.

Some interviewees suggested not to bother with resistors because it is energy depleting. Kim suggested, "If they're just a dyed in the wool believer and there's no way anything you say will make a difference, save your energy." Clariana is wary about who she engages with and is mindful about protecting herself, "Especially when people are just blatantly racist, or rude, or... degrading... There's a balance between someone who's willing to have a conversation and listen and people who are just rude. Sometimes it's not worth the fight in terms of your own mental health." Alex also thinks about it from a self-care perspective, "The biggest risk factor to yourself, is that it can demotivate you... it takes from you if you spend a lot of time trying to change that one very resistant person." Alicia understands that everyone has different opinions and on this basis she chooses not to expend energy in that direction:

> It's very draining. I know I have limited effort that I want to put into something, and I'd rather use it to work with people who want to work with me than waste on those who I'm trying to convince. It's either they have that in them or they don't, and it's something I've learned to accept... Either somebody has a heart for giving or they don't, and you don't need everyone to be with you or on the same page as you.

Putima indicated that perhaps the negative reaction comes from a place of fear:

> It took me a long time to realise you can't change people's minds – if they don't want to... if they're very cemented in their views, and they might be surrounded by the people like them... they can always justify or live with the stuff that they do... And I think there's a lot of fear... that they're going to lose... possessions, or power, or influence.

Delphine also chooses not to engage but her reasons are about respecting the right of someone to have different opinions:

> The one tactic I know is what I learned from the Bible and that is I have to change the way I respond to their response – they have their choice to make and if that's what it boils down to – people have their own choice and that has to be respected.

Adrian's default strategy is for people to "take it or leave it – No way of geeing them up except me walking my talk" because he accepts that "eventually someone else with more influence will influence you, or you will remain as you are".

It may be sensible and logical to recognise how hard it is to change someone's mind and that the quest may be futile. However, Elise is mindful of finding a balance between not wanting to remain in a social circle that agrees with everything she says and wasting resources on people who may not have the same degree of openness:

> When you stay in your own silo it's easy for things to become a bit of an echo chamber, but then if you try and really engage with people on the other side of the fence sometimes, like sometimes people have just such fundamentally different belief systems or

views that it almost feels like it's impossible to change their mind just as it would be impossible for them to change my mind. So that's not something that I have an easy solution for.

Similarly, Ben struggles with finding an appropriate balance:

I've learned over the last five or six years to try and resist spending too much time [engaging with denialists]. It's hard though... I'm still guilty, as I suppose many people are, of dedicating too much time to focusing on people that I think I can change their minds... What's the point wasting my time on people like this, where I could be dedicating my energy to people who actually are engaging in these ideas in good faith and might actually be motivated to make a positive difference in the world?

Listen

Depending on the role and sphere of activity, activists may have a strategic responsibility to listen to and understand the arguments and motivations of resistors in order to use this information. Some interviewees said that taking the time to understand more about what the denialist was thinking was critical for formulating an appropriate strategy for engagement. Depending on the degree of extreme thinking, according to some interviewees, listening was the catalyst for potentially triggering other approaches. They wanted to understand the fears of denialists, what they thought they would be losing, and what they believed would be changing. The interviewees talked about *reading* people. They said they valued diverse perspectives and genuinely wanted to hear their opponents' views. Understanding their argument meant listening intently and probing for more information. They suggested that the first step was recognising the complexity in any issue, and then keeping an open mind to explore where the person's opinions were coming from.

Mike sees these interactions as an opportunity to self-reflect and seriously consider the person's argument; he does not want to be so close minded as to become a denialist on the "other side". Khalil expresses how much pain it causes him to listen, but he does it anyway:

My strategy is to think I need to talk to them and listen to them. Sometimes I go to Fox News, sometimes I can't stand it anymore, but I want to see how they see things. Even if I hear something and I'm upset, I keep listening to understand their argument. So, you can understand how they think, how they're manipulated, how they manipulate other people.

Ricki also acknowledges that the process of listening is not easy but essential for understanding other people's perspectives. She said:

> When you're working in activist spaces, you need to understand your opponents very well... You've got to understand your audience. If you want to genuinely empower change, you have to put yourself out into that uncomfortable space and learn about allyship, learn about opponents, learn about the language that you are not bringing to the table. What is the dominant view of people?

Clariana notes, "I'd have conversations to look into where the actual barriers stem from... There are so many options and hooks in conversations that we ignore!"

Marilee worked in a support role in a political party that did not align with her political beliefs. She found this to be a great learning experience as the role provided her with tactful skills:

> It helped me see that people with other views, there's nothing wrong with them; intelligent people can disagree. It opened my eyes to see that they're not wildly fascist and hateful, limited minded, small minded people, they just believe different things than me and have different values than I have... I think when you really listen to other people then they do a better job of listening to you.

Noraini spends time developing her skills to understand the motivations of people who have different views to her own. By questioning and finding out more, she finds herself in a position to share her own point of view, which sometimes provides opportunities for changes in thinking. She said:

> It became clearer to me how to understand people and what drives them and how to motivate them. I began to understand how my mind and their mind worked. Perhaps not to change their mind immediately, but at least to shake their mind. Ask enough questions to have them start questioning themselves, that's already a win. Pacing and leading. You understand what they're doing and once you shake that a bit and they're comfortable and know that they can trust you, then they follow suit easier, so that's my tactic.

This aligns with Jody's tactics for engaging with resistors:

> It's about knowing your audience as well and understanding that... my perspective is not always the right way, so valuing what they have

to say and acknowledging, recognising that there can be many differences, many opinions, and… listening to what their myths – or what we may perceive as myths – but they're only working with information what they have. And it's respecting as well, respecting what they have to say and their values or their opinions.

Listening and authentically questioning is a theme that consistently appears in literature promoting the benefits of friendship and sharing how to make powerful connections (Owen, 2022). Some key strategies whilst conversing include being present, staying in the moment without multitasking, believing you have something to learn, setting aside your opinion, and not just listening to come up with a reply, but with the intent to understand (Headlee, 2015; Parker-Pope, 2019; Scott, 2021). Journalist Kate Murphy (2020) in her book *You're not listening – What you're missing and why it matters* highlights the situations in modern life where people are not listening to each other and relates this to our feelings of isolation, loneliness, and intolerance. She makes the connection between our lack of collective listening skills and the effects that this has had on a political, social, and cultural level:

> Listening is something you do or don't do every day. While you might take listening for granted, how well you listen, to whom, and under what circumstances determines your life's course—for good or ill… And to listen poorly, selectively, or not at all is to limit your understanding of the world and deprive yourself of becoming the best you can be.
>
> (Murphy, 2020, p. 4)

Connect

Interviewees suggested that sometimes they would listen to the resistor but make it clear that they did not agree. They let the resistor know that they were open to continuing a conversation. They did not force their opinion on the resistor but still thought it worthy to speak up in a non-confrontational manner. In her softly spoken, gentle way Vijeta recommended:

> Look at the reactions people give you. Sometimes I back off a bit and then I realised its ok – I can make my point. I'm not responsible for their choices and beliefs. Say your piece but don't try to take responsibility for other people. Don't let that stop you saying what you need to say… some people come with strong opinions, and I need to challenge them sometimes.

Alex suggested that ensuring her interactions with denialists are based on hope provides her with a solid foundation for engagement. She gave an example of someone who questioned her about recycling as they were concerned about the large volume of material still going into landfill. Alex responded by sharing how she believed her actions were connected to a larger resistance movement. She did not let the interaction affect her own source of strength. Instead, the interaction resulted in reinforcing her underlying motivations:

> If we keep the habit and we show that we've got a full recycling bin every week then hopefully that shows the council that we actually do care and that this is something that's important to us... it's important to maintain habits because they are a form of activism within themselves... Try to address [the questioning] from a sense of hope and a good place, rather than engaging in an argument or using a lot of your energy on something that potentially won't change. That is the biggest risk factor to yourself – it can demotivate you – it takes from you if you spend a lot of time trying to change that one very resistant person.

Jenny looks for things that she has in common with people in denial. She said:

> Bridge that dialogue and have a dialogue that's meaningful. Stay calm and not aggressive. If you're yelling at someone, that's not helpful. People are going to be really defensive if you're attacking them. You can't just say someone is a bad person because there's something you disagree with. I always bring conversations back to something we have in common and have that connection as the last thing I leave them with.

Interviewees suggested that when it becomes apparent that values or beliefs around an issue are not in alignment, these moments can provide opportunities for gentle and respectful ways of opening a conversation. In a situation where Elise was watching television with a member of her extended family, the relative complained about a political issue. Elise said, in as friendly a manner as possible, "criticising from an armchair did not serve anyone". Elise was surprised at the effect of this interaction, "I didn't realise that it had stuck with her until a few days later when she brought it up with me and asked me about some of the campaigning work I was doing. Then I got her involved in this letter writing piece." Elise previously felt frustrated when people would complain, and she did not say anything in response. Following this positive outcome, she has subsequently channelled this frustration by trying similar strategies in scenarios as they arise.

Alex and Talia take a nuanced approach depending upon where people are on the continuum of change. At a conference, Alex learnt about the different levels of readiness for change (e.g., Prochaska & DiClemente, 1983). She explained: "[Some people] are not going to accept the ideas... they're not worth your time. They're a lot of effort and they won't change, so you focus on [people] willing to have a conversation." Talia similarly wants to work with people who are more receptive, "Deniers are often very strong headed and hard to shift, so I try and work on those people who are more in the middle and undecided."

In these types of circumstances, the interviewees emphasised the importance of face-to-face interactions. To bring down barriers around gender in sport among the families of young people in a local competition, Domuto talked in a one-on-one friendly way about the benefits of participating and emphasised the importance of compromise. He said, "Many believe playing ball is only for boys. Often fathers challenged me, but I convinced them and they finally agreed to let their daughters play." Darrell also developed his strategies for dealing with people who were obstructing his advocacy work. He would always try to set up opportunities to talk in person and it involved a calm and considered approach that was careful not to inflame emotional responses.

Although Marion believes that most social change only occurs when large numbers of people genuinely want change to occur, and acknowledges that individuals have a limited capacity to influence this process on a societal level, she does recognise that individuals can influence people with whom they have a meaningful relationship. Marion suggested that interpersonal connections may not lead to change, but change is unlikely to happen without meaningful interactions: "the establishment of an authentic and meaningful connection between people... social connection is crucial to bring about change." She has had her opinions changed by people she respects so she uses the platform of respectful relationships as the foundation for engaging. She shares her practical approach:

> Sometimes I'll spend a whole day with people in the witness facility and you'll be talking to them sometimes for many hours. And the conversations come up about politics and about climate – and you can be influencing people quite gently. And they've already got a lot of trust in you because you're doing these really important roles with them and supporting them... If people are really quite actively against what you're saying, then that's more difficult... I

do challenge them gently about things and try and change their views. If they were open to receiving information, I might send them an article or two about something that they're feeling negative about.

Talia, Alex, and Marion's approaches of working from a place where there is trust and an emotional bond, aligns with recent evidence from the behavioural sciences. The Australian environmental organisation Climate for Change (2023) uses the power of behavioural science by harnessing the conversations that occur in trusted social networks. They divide their audience into six categories: opposing, cautious, wavering, passive, active, or leaders. They then specifically focus their initiatives on three of the middle categories: "wavering (say they support strong action from our leaders, but often waver when faced with real-life examples), passive (support strong climate action, but do not take action towards it) and active (take action occasionally, often when asked)" (p. 7). Using a party plan model, they encourage people to talk about climate change within their networks and support each other to take action. Social research data in their impact report demonstrates significant behavioural change continues three months after the conversations (Climate for Change, 2023).

Rebecca Prince-Ruiz and Joanna Atherfold Finn (2020) share their journey in their book *Plastic free – The inspiring story of a global environmental movement and why it matters*. From the outset of their movement, which started with one person, they have avoided asking, "What difference can one person make?" and focused on "How can we continue to create change together?" (p. x). The small team organically expanded throughout their networks as other people realised the value of their approach. It has since evolved into a global movement that helps millions of people engage in strengths-based practical change (Plastic Free Foundation, 2023).

Social scientist Rebecca Huntley (2021), in the book *How to talk about climate change*, suggests that there are people who are alarmed and engaged with the issue of climate change and others who may be concerned but who choose to disengage. Some may be sceptical while others are active deniers. She writes that these different levels are all connected to our underlying emotional reactions – anxiousness, fear, anger, or detachment. Understanding these emotional connections can help with coping on a personal level and can unlock what it would take to convince another person to act. Acknowledging we are human beings who do not readily change our behaviours based on new factual knowledge, Huntley (2021) writes:

Reason and emotion. Imaginative and reckless. Constantly evolving and selfish. Loyal to tribe above almost all else. Any approach
to communicating about climate change that doesn't take these
human characteristics into account will not be very effective.

(p. 16)

To be motivated into action, Huntley recommends that new information must connect with our lives, relate to our social group, and be
meaningful, digestible, and actionable.

Challenge

Although most interviewees did not recommend engaging with resistors, some interviewees thought that it was important, almost their
duty, to not let some things go unchallenged. Khalil said:

> I'm an advocate of talking to people who resist or disagree. Other-
> wise, you're hiding them from your eyes, but they still exist. And they
> are people who vote and who live in your community. Most of the
> time it's very frustrating. There's so much evidence but they won't
> listen to any scientists. You can't ignore them and pretend they don't
> exist. You have to talk to people, especially who disagree with you.

Phyo Phae Thida engages with all types of people because she feels
obligated to share her information with the wider community. Sharing
with her family is an absolute minimum, but to progress social change,
she recognises that she has to join with other people – that means
raising awareness. She acknowledges how much is at stake and wants
to do everything she can to promote change:

> If I want to change something I need to speak out. It doesn't really
> matter if they listen, you have to speak out... The first time they
> might not listen but if you keep talking and they keep hearing the
> same message again then maybe somehow they will think that they
> want to join you. Even if they don't change, at least they're
> aware... I'm not forcing "oh you must come and protest with me",
> but when we have a common interest, they will offer if they want
> to join. We invite them, but not forcing. The first thing is talking
> and sharing our interest and our perspective on certain things then
> some people will feel the same and will join me. Some people who
> have a bit of a different opinion and perspectives, they might also
> join me to have the discussion even if they disagree with me.

Research suggests that it may be worthwhile to speak up in situations where you hear a piece of misinformation because humans routinely rely on what other people within our social networks say to justify our beliefs (De Cruz, 2020). Philosophers Giulia Terzian and María Inés Corbalán (2021) argue that in these times of rising misinformation there are some topics where we ought to speak up. Depending on the circumstances, they encourage people to engage when false beliefs are likely to lead to harmful or hateful situations (Terzian & Corbalán, 2021). Unless you are going to be completely ignored or harmed physically, it might be worth the courage and patience it requires to have direct and respectful conversations with people in your close social circle. Not interacting at all may make things worse by emboldening their beliefs and attitudes because it hides the fact that there are differences in views.

Professor of philosophy Matt Ferkany (2022) argues that there is a place for providing information and pointing out inconsistencies. He prompts readers to think about whether a potential breakdown in the relationship is worse than resenting and complaining about the person in a non-respectful way behind their back. Ferkany (2022) acknowledges that direct, respectful, and spirited exchanges may have to occur many times before change happens:

> Arguing about the kind of world we want to create is how social progress is made, and we have a responsibility to contribute to this where we can. We are uniquely positioned to do this with family and friends, with whom we bear a special bond. Because of this bond, we might be able to re-engage in difficult conversations with them, perhaps many times over many years... Sowing doubt, prompting more careful thought, and raising awareness about other possibilities can all be valuable in themselves and might even lead to bigger changes later.
>
> (p. 1)

At the very least, if someone is wavering on the precipice of resistance, they will know that they can come to you for alternative information and support (Badham, 2021).

Reflection

Activists will always encounter people who disagree or who are in denial. We suggest that it is worthwhile for activists to claim their stance and use interpersonal strategies to help change the attitudes and behaviours of people within their trusted social circles. A starting point may be acknowledging that engaging with difficult audiences can be worthwhile: "being present and rebutting science denial still makes a positive difference" (Schmid & Betsch, 2019, p. 935). We can learn from conversations with people who hold different views. We can use the information to review our positions and refine our plans. We can be open to modifying or qualifying our opinions based on new information. Using strategies such as those outlined in this chapter can increase the likelihood of having an effective interaction and avoid wasting time and resources focusing on behaviours that will have a very small or potentially negative impact (Behaviour Works Australia, 2023).

An example comes from academic Lee McIntyre (2021) who argues about the importance of taking on people who deny science in his book *How to talk to a science denier – Conversations with flat earthers, climate deniers, and others who defy reason*. He shares his experiences in converting people who believe the Earth is flat, discussing climate change with coal miners, and chatting with friends about genetically modified foods. He engages them in a conversation on what is special about science in a calm and respectful way. He emphasises how positive changes can arise from relationships based on trust, respect, warmth, and connection. World champion debater Bo Seo (2022) also shares his secrets about how to gather information, find truth, speak with lucidity, organise an argument, and be persuasive in his book *Good arguments – What the art of debating can teach us about listening better and disagreeing well*. His approach is to avoid malicious or personal attacks and maintain the relationship with the person (Seo, 2022).

The final strategy we present comes from forensic psychologists Emily Alison and Laurence Alison (2020). Their area of expertise is in terrorist interrogations and through their challenging experiences they have developed a simple model of interpersonal communication. Their book *Rapport – The four ways to read people* emphasises that developing rapport is a skill that can be learned through practice. Their research presents a model based on the understanding that everyone has an interpersonal comfort zone. Some people gravitate towards conflict, whereas others may want to capitulate when uncomfortable. Some people want to be in control, while some look for cooperative ways of engaging. Identifying your natural engagement style is key to avoiding default negative behaviours. It is then beneficial to learn to cultivate more unfamiliar styles of interaction so that you can tailor your behaviour and deal more effectively with diverse situations.

Alison and Alison (2020) suggest that to move away from cultural tribalism and bridge the divide between the polarised views that are evident across society, we need to understand more about someone's core beliefs and values even when we disagree. Honesty, empathy, and autonomy are required to build a solid foundation for positive interactions, effective communication, and meaningful relationships. They suggest the following prompts can help us to hone our ability to modify interactions for mutually beneficial outcomes:

- Am I being honest or am I trying to manipulate the other person?
- Am I being empathic and seeing things from their perspective or just concentrating on my own point of view?
- Am I respecting and reinforcing their autonomy and right to choose, or am I trying to force them to do what I want?
- Am I listening carefully and reflecting to show a deeper understanding and build intimacy and connection? (Alison & Alison, 2020, p. 146)

Conclusion

Dealing with people who disagree with us or who are in denial takes courage. Even listening to people who think differently can be difficult. Therefore, any action to speak up about a belief, or a political, or social view inevitably requires a considerable amount of intellectual and moral courage (Rabieh, 2006). Steadfastly holding onto views can be risky as it involves a certain degree of self-sacrifice, but it can also be impressive and compelling when it involves facing up to challenging situations: taking on the risk, making a self-sacrifice, and then accomplishing the goal (Rabieh, 2006).

However, it is common to have a situation where the everyday activist and the opposer can both think they are right. Impasses in conversations are likely if both parties, who each believe the other is in denial, are steadfastly, seemingly courageously, holding onto their views. Many people are unwilling to listen to opinions that are not in alignment with their own. To develop a tolerant society, we need to reflect on our personal ability for listening to diverse opinions without getting defensive. We need to build our skills for addressing arguments without disengaging:

> Perhaps the most useful courage that we have as individuals is the courage to be able to scrutinise our own opinions... The courage to be open to hearing and wrestling with the fact that our opinions are wrong because, at least according to Socrates, who claims that

it's a greater good to be refuted than to refute, it's the only path to clarity or what we call human wisdom.

(Rabieh & Boag, 2022, p. 112)

References

Alison, E., & Alison, L. (2020). *Rapport: The four ways to read people.* Penguin.

Badham, V. (2021). *QAnon and on a short and shocking history of internet conspiracy cults.* Hardie Grant Books.

Behaviour Works Australia. (2023). *Behaviour Works Australia.* Monash University Sustainable Development Institute. https://www.behaviourworksaustralia.org/

Carroll, S. B. (2020). The denialist playbook. *Scientific American.* https://www.scientificamerican.com/article/the-denialist-playbook/

Climate for Change. (2023). *Climate for Change.* https://www.climateforchange.org.au/socialresearch

De Cruz, H. (2020). Believing to belong: Addressing the novice-expert problem in polarized scientific communication. *Social Epistemology*, 34(5), 440–452. doi: https://doi.org/10.1080/02691728.2020.1739778

Farrell, J., McConnell, K., & Brulle, R. (2019). Evidence-based strategies to combat scientific misinformation. *Nature Climate Change*, 9(3), 191–195. doi: https://doi.org/10.1038/s41558-018-0368-6

Ferkany, M. (2022). Why you shouldn't shrink from challenging your loved ones' views. *Psyche.* https://psyche.co/ideas/why-you-shouldnt-shrink-from-challenging-your-loved-ones-views

Headlee, C. (2015). 10 ways to have a better conversation. TedTalk. https://www.ted.com/talks/celeste_headlee_10_ways_to_have_a_better_conversation

Huntley, R. (2021). *How to talk about climate change in a way that makes a difference.* Murdoch Books.

Kahn-Harris, K. (2018). *Denial: The unspeakable truth.* Notting Hill Editions.

Macquarie Dictionary. (2016). *Macquarie dictionary online.* Macquarie Dictionary Publishers, Pan Macmillan Australia. www.macquariedictionary.com.au.

McIntyre, L. (2021). *How to talk to a science denier: Conversations with flat earthers, climate deniers, and others who defy reason.* MIT Press.

Murphy, K. (2020). *You're not listening: What you're missing and why it matters.* Macmillan Publishers.

Nadler, S., & Shapiro, L. (2021). *When bad thinking happens to good people: How philosophy can save us from ourselves.* Princeton University Press.

Oreskes, N. (2019). *Why trust science?* Princeton University Press.

Owen, M. M. (2022). The art of listening. *Aeon.* https://aeon.co/essays/the-psychologist-carl-rogers-and-the-art-of-active-listening

Parker-Pope, T. (2019). How to be a better friend. *The New York Times.* https://www.nytimes.com/guides/smarterliving/how-to-be-a-better-friend

Plastic Free Foundation. (2023). *Plastic Free* July. https://www.plasticfreejuly.org/

Prochaska, J. O., & DiClemente, C. C. (1983). Stages and processes of self-change of smoking: Toward an integrative model of change. *Journal of Consulting and Clinical Psychology*, 51(3), 390–395. doi: https://doi.org/10.1037/0022-006X.51.3.390

Prince-Ruiz, R., & Finn, J. A. (2020). *Plastic free: The inspiring story of a global environmental movement and why it matters.* New South Wales Press.

Rabieh, L. (2006). *Plato and the virtue of courage.* Johns Hopkins University Press.

Rabieh, L., & Boag, Z. (2022). A political virtue. *New Philosopher*, 38(Courage), 106–112.

Schmid, P., & Betsch, C. (2019). Effective strategies for rebutting science denialism in public discussions. *Nature Human Behaviour*, 3(9), 931–939. doi: https://doi.org/10.1038/s41562-019-0632-4

Scott, E. (2021). Strengthen friendships with good listening skills. *Very Well Mind.* https://www.verywellmind.com/strengthen-your-friendships-with-good-listening-skills-3144970

Seo, B. (2022). *Good arguments: What the art of debating can teach us about listening better and disagreeing well.* Scribner Australia.

Spector, M. (2009). *Denialism: How irrational thinking hinders scientific progress, harms the planet, and threatens our lives.* Penguin.

Terzian, G., & Corbalán, M. I. (2021). Our epistemic duties in scenarios of vaccine mistrust. *International Journal of Philosophical Studies*, 29(4), 613–640. doi: https://doi.org/10.1080/09672559.2021.1997399

7   Everyday evaluation

Effective activism is predicated on one question: *Am I making a difference?* This chapter provides practical strategies based on informal everyday evaluation. Drawing insights from interviewees, the chapter discusses the elements of reflection, tracking, and sharing. It provides practical guidance for identifying strengths and weaknesses; considering the worth, merit, and significance of activist efforts; and determining their effectiveness. Evidence can be used to improve activist efforts, learn, motivate, influence, advocate, recruit followers, and engage donors. This chapter shares a roadmap for collecting relevant data, evaluating usefulness, transforming information into compelling and concise formats, enhancing the reach of evidence through strategic sharing, and elevating impact by incorporating insights into future actions.

Most interviewees, when asked about their engagement in monitoring and reflection, generally noted that they neglected such practices. Either tracking efforts were completely absent or they felt that what they were doing was inadequate. Some had tried, but the efforts had been half-hearted. Others planned to evaluate but found it difficult to get started. Many interviewees expressed regrets over their non-engagement with monitoring and evaluation. They thought it was too complex or did not know how to perform an assessment. Other reasons included insufficient resources – notably, a lack of time. A recurrent theme among the interviewees was their perpetual state of urgency. This was characterised by a consistent cycle of either responding to immediate crises, recuperating from the crises, or planning for the next stage. Some activists were simply content with making positive assumptions about outcomes. Darrell candidly captured a common sentiment: "It was totally voluntary, it was intense... There was no time to complete excessive records, no time for that. You just put them in a box... And after a few years, I'd throw them out."

Despite interviewees generally struggling to articulate their approach to reflecting, tracking, and knowledge sharing, a subset of interviewees

DOI: 10.4324/9781003333982-7

imparted strategies that have worked for them. These strategies include the establishment of monitoring frameworks, using key performance indicators, and conducting regular self-assessments of their work. Additionally, interviewees recounted recording testimonials, keeping track of social media metrics, verbally asking for feedback, asking for experts to review their work, graphing sales figures, collating anecdotal evidence, monitoring their membership base, undertaking polls, presenting at conferences, recording stories, writing annual plans and reports, collecting feedback from letters and emails, and tracking their participants' progress over time.

This chapter is based on actions that the everyday activists interviewed for this book use in practice, ideas that they would like to implement, and problems that they requested support to understand and solve. While centred around interviewees' narratives, this chapter draws strongly from relevant evaluation literature. We have avoided using technical evaluation jargon. Instead, we talk about monitoring, reflection, and knowledge sharing – referring to these as *everyday evaluation*, as outlined by Australian sociologist and evaluator Yoland Wadsworth (2011) in her book *Everyday evaluation on the run*. We draw upon previous research undertaken with small community-based organisations that found the vast majority of program and activity improvement occurs through small-scale reflective practices, conversations between personnel and recipients, and simple tracking of available data (Kelly, 2019, 2021, 2022).

You may think you are not evaluating, but you are evaluating all the time (Kelly & Rogers, 2022; Wadsworth, 2011). However, this type of everyday evaluation is rarely identified or highlighted for its value toward improvement and learning due to its informal and continuous nature. Everyday evaluation manifests in myriad ways: discussing strategies with other activists, holding check-ins with community members, thinking about the successfulness of an activity on the journey home, or looking at the metrics from an event. It is about reflecting, assessing, and making judgments about effectiveness through examining efforts, outcomes, and strategies to understand whether goals are being achieved and if adjustments are needed (Wadsworth, 2011). It provides real-time feedback using monitoring data, feedback from participants, and evidence from practice to fuel critical reflection.

Evaluation as a catalyst for social change

Activists can leverage evaluation to amplify and interrogate their efforts. Before activists even begin to address their cause, they can assess salient information, such as research findings, to better understand the needs, articulate the *why*, and guide development of actions (Kelly & Rogers, 2022). This evidence can help determine whether you may be doing more harm than good. It can predict and explain the logic between actions and expected outcomes and justify decisions surrounding resource allocation and approaches. Further, engaging in a well-considered process before leaping to the *doing* can support identification of what should be monitored and evaluated to ensure continuous adaptation according to good practice and changing trends.

Evaluation identifies what is effective and what needs improvement, enhancing both accountability and credibility (Patton, 2018). An evidence-based approach informs strategic decision-making and ensures that actions align with desired outcomes. Activists can then better allocate resources, refine strategies, and maximise influence. Evidence from evaluations can also attract funding. Demonstrating tangible impact can boost an activist's ability to secure resources and extend their initiatives. By showcasing the problem and highlighting effective solutions, evaluation can be a powerful tool to draw support, donations, and involvement.

The dynamic nature of activism requires continuous adaptation. Evaluation offers a platform for reflection, helping activists evolve their methods, adapt to new challenges, and maintain the relevance of their campaigns. Monitoring shifts in context helps activists adjust strategies, ensuring their efforts remain relevant and enduring.

Engaging stakeholders in evaluation creates collective ownership of the change-making process. Collaborative approaches enhance sustainability and foster a shared sense of purpose. As previously discussed in Chapter Five on cooperation, building stakeholder buy-in and celebrating collective achievements sustains momentum and furthers collaboration.

Evaluation itself can be a valuable tool for advocacy and activism (House, 1990, 2007). Well-documented successes and lessons learned, validated through evaluation, become potent narratives that everyday activists can disseminate to broader audiences. These narratives can bolster advocacy efforts, facilitate policy change, amass support, and catalyse wider participation in the activist movement (Fetterman, 2003).

Interviewee and professional evaluator Khalil passionately feels that evaluators have a duty of care to highlight social justice issues in their analysis and use this information for advocacy and activism. He explained that "evaluative thinking is about trying to help, trying to advise to do the right things, because you have evidence when you do evaluation". Further, evaluating using an equity and social justice lens can help everyday activists consider other angles of their contribution and highlight areas for self-improvement. This includes considering decolonisation and adopting participatory evaluative approaches that purposefully seek to redistribute power, reconsider dominant frameworks, and incorporate the voices of marginalised communities that activist issues affect (Fetterman, 2023; Kelly & Htwe, 2023). In sum, evaluation equips activism with a strategic edge that can enhance reflectiveness, equity, and inclusion. Adopting the evaluation techniques detailed in this chapter allows activists to amplify their positive impact on the world.

Monitoring or tracking

Monitoring might sound complex, but it need not be. It refers to the systematic and ongoing process of tracking and collecting relevant data to gauge whether the intended objectives are being achieved and if any adjustments are required (Kelly & Reid, 2020). Monitoring provides real-time insights into the performance of activist efforts and informs decision-making.

Many interviewees highlighted that they regularly gather monitoring data, and some have set up systems to track their progress and be more systematic in their collection and storage. They developed indicators that helped them measure change over time. Some shared how they used the initial business plan as a benchmark to track change on an annual basis. Others set annual objectives, which they broke down into tasks.

Monitoring can be as simple as a spreadsheet that records dates and participant numbers, or it can be more comprehensive, as detailed in the steps below. Checklists have been created by Western Michigan University's evaluation centre and the Better Evaluation website has a range of useful tools (Global Evaluation Initiative, 2023; Western Michigan University, 2023). The following guidance is a simplified version of a more conventional approach, such as that outlined in Markiewicz and Patrick's (2016) book on *Developing monitoring and evaluation frameworks*. These suggestions can be adapted to each context and extended or pared down depending on need and resources available.

Establish goals

A good place to start is by clarifying your overarching goal. What is the change you want to see? For example, the goal of a community garden might be that people are healthy and socially connected in their local community. Then think about the sub-goals or outcomes that contribute toward your goal. Precise achievable outcomes serve as a basis for monitoring progress and evaluating success. This might include things like: 1) local people have access to nutritious, fresh vegetables and 2) local people have a place to meet and interact. A project timeline can break this information into tasks, milestones, or checkpoints.

Set indicators

Determining indicators that align with your outcomes can help measure the impact you intend to create. Thinking about the information you need to show that you have achieved your outcomes can help with setting indicators. Outlining how, where, when, and who will collect this data helps clarify an implementation plan for monitoring. Indicators for the first outcome illustrated above could include items such as the weight and variety of vegetables grown and harvested.

Gather information

Once you know what it is you want to measure, you can gather data that is relevant to your indicators. This could include quantitative data (e.g., number of participants, funds raised, kilograms of vegetables grown) and qualitative data (e.g., participant feedback, media coverage, anecdotes/stories). Roy, Talia, Tanya, and other everyday activists track their donations. Jamal, Phyo Phae Thida, Jenny, and others track their reach, reposts, and followers on social media. Mehreen, Caroline, and Matt track the number of kilograms of waste their project diverts from landfill.

Many everyday activists we interviewed for this book mentioned the value of longitudinal data where they were able to gauge long-term impacts and assess sustainability. Mehreen, Jenny, Alicia, Aruna, and Nafiz checked in with people they had assisted years before and saw positive progress. Alex, Nicole, and Susanne's initiatives were continued by others after they stepped away and have gone from strength to strength. Vijeta received communication from now-adult women she had worked with to prevent child marriage who credited her program with supporting them to reject their families' underage marriage expectations.

Setting benchmarks, developing standards, or finding ways to track change over time can help compare current results with past performance or industry best practices. These can be simple and even based on anecdotal observations. They can be written into stories of change. For example, Kim mentioned that her anti-duck shooting posts were previously removed from online bird watching and photography groups, but now they are allowed to remain and garner attention. Domuto tracks the extension of his sports programs into other regions and counts interactions with a growing number of groups. Tanya collects and measures progress by comparing previous years' donations and memberships against the current year. Caroline demonstrates how much money they save each month in her workplace recycling program.

Adapt and improve

Another aspect of everyday evaluation includes using monitoring data to adapt and adjust strategies. Flexibility and learning from evidence are key to refining approaches and optimising impact. Making time for reflective practice creates a space to consider and action these changes. A simple way to ensure uptake of reflections, lessons, and findings is with an action plan that outlines what needs to be done, who will do it, how, and when. Having an action plan, task list, or other method of tracking the uptake of findings and recommendations helps ensure that this evidence is invested back into the activist endeavour.

Share findings

By integrating these practical strategies into daily routines, everyday activists can effectively track the impact of their initiatives. Monitoring not only informs adjustments for enhanced effectiveness but also provides evidence to gain support, influence decision-making, and spread awareness. Most importantly, examining monitoring data demonstrates evidence of progression toward goals, which can be used to acknowledge and celebrate milestones and successes. Ben notes that, "Particularly with advocacy, you don't necessarily see the rewards of your work straight away. It can sometimes take a really long time for change to happen. So, I do treasure those moments and value them." As such, monitoring provides opportunities to celebrate small wins, bolster morale and motivation, and invigorate your activist community.

Reflect

Reflective practice supports everyday activists to evaluate and enhance their approaches. It can be conducted individually, collaboratively among peers, or in conjunction with community members. Reflection can facilitate critical analysis, prompt self-awareness, and assist with understanding the dynamics. It is important to cultivate an environment where the work can be safely interrogated and failure can be accepted. As highlighted by Nafiz, learning from mistakes is key to improvement and effectiveness: "Actually, that simple sentence... *that didn't work*, contributes to the next cause, and next intervention." Some interviewees even took time to consider whether they were doing the right thing at all or doing more harm than good. Kim mentioned that it is important to think at a high-level about activist work. She explained that leaping to solutions without adequately unpacking the problems and reflecting on the likely consequences of potential actions can create additional problems and fail to address the initial impetus for the action. Reflecting on the overarching problem and solutions can help assess direction and adjust efforts as deemed appropriate.

Individually, reflective practice provides an avenue for self-examination and self-appraisal. By engaging in solitary introspection, activists unpack the nuances of their actions, experiences, and outcomes. This fosters an intimate understanding of personal motivations, biases, and areas for improvement. Additionally, it stimulates a culture of continuous learning, enabling activists to iteratively refine their strategies based on their evolving insights.

Interviewees mentioned that they take time for self-reflection and track their personal development. They engage in conversations with others about their progress and approaches. They reflect on their behaviour and the behaviour of others to understand what they could learn and how they could grow. They said that formal and informal reflective practices helped them identify what changes had happened, what changes they wanted to happen, and where they could be more effective. Tamsin explained her structured approach to introspection where she uses her Saturday morning walk to reflect on her internalised biases, unpack narratives that she has accepted as fact, and consider ways that she can do better. She found tools such as completing weekly diary prompts in Layla Saad's (2020) anti-racism book *Me and white supremacy* incredibly helpful to guide her thoughts.

Among peers, collaborative reflective practice provides activists with a platform to engage in open dialogue, share diverse perspectives, and solicit constructive feedback. Through the lens of others' observations, everyday activists gain fresh insights into their approaches, and expose latent blind spots. This collaborative engagement fosters an environment of collective growth, wherein the synthesis of varied viewpoints informs strategic recalibrations, enhancing the overall effectiveness. Kim noted the value of the activity:

> When I was immersed in the quarry stuff, there wasn't much time for reflection. But coming out of it, it was great to stop and have a look at what we'd done and analyse it a bit. During the hiatus between two proposals for the quarry we had to evaluate what we'd done and how we'd do better. We did that reflection through lots of meetings.

Ricki highlighted the great potential of collaborative time to engage in reflective discussions: "we can share ideas, we can chat, and that's how you build your community apparatus, through likeminded people who want to share and learn from each other, and support each other, and that's how community activism grows."

Engaging community members and recipients of activist projects in reflective practice aligns with participatory principles, promoting inclusivity, and democratising the evaluation process. Speaking to community members captures lived experiences and contextual nuances that might otherwise be overlooked. The insights gleaned from community interactions enable adjustments that resonate with the community's needs and aspirations.

Making space for reflection in whatever way works best for your situation is key. Reflective practice can simply involve making space for thoughtful conversations and recording the key points or it can help to have questions on hand to prompt discussion (Schon, 1983). These can be a set of generic questions that are used each time or could be informally devised on the spot based on the situation. They can be used to evaluate an event or action, or to interrogate an entire approach to an activist cause. Gibbs' (1988) reflective practice cycle can guide dialogue by asking questions around the following stages: description, feelings, evaluation, analysis, conclusion, and action plan. Reflecting through this cycle can provide a comprehensive framework for evaluating and enhancing activist activities and approaches. These are example questions using Gibbs' framework to guide a reflective practice session:

Description

- What specific event or activity am I reflecting on?
- Who were the individuals involved in this event, and what roles did they play?
- What were the intended outcomes or objectives of this event?

Feelings

- What were my emotional responses during this event?
- How did my emotions influence my perceptions and actions?
- Were there any particular feelings that stood out, and why?

Evaluation

- What aspects of this event do I view as successful or positive?
- Are there any elements that I consider challenging or less effective?
- How well did this event align with my personal values and activist goals?

Analysis

- What factors contributed to the outcomes observed during this event?
- Are there relevant theories or concepts that could help unpack what occurred and why?

Conclusion

- What key insights have I gained from evaluating this event?
- How could I have mitigated any negative aspects of the event?
- What lessons can I draw from this event to enhance my future actions as an activist?

Action plan

- Based on my reflections, what concrete actions can I take to improve future events?
- How can I leverage the strengths and successes of this event to enhance future impact?
- What specific steps can I outline to align my future activities more closely with my activist objectives?

Translating evidence

Rather than simply accumulating knowledge, it is important to effectively transform it into actionable insights. By translating complex data into relatable narratives, activists can share knowledge with diverse audiences. Thus, *knowledge translation* reframes sharing as a form of activism, as it equips others with the tools to effect change.

Knowledge translation is the process of transforming or synthesising information into easily digestible material that is more likely to resonate with target audiences. It entails accessing, creating, and disseminating knowledge for the purpose of making decisions and generating action (Rogers & Malla, 2019). Information can then be communicated to diverse audiences to foster informed engagement and collaboration. This helps bridge the gap between data and real-world impact. It can help activists determine whether there is evidence to support their course of action, empower stakeholders to make informed decisions, and influence others to get involved with the cause.

Key elements of knowledge translation are simplifying, tailoring, and utilising diverse platforms to share the same message in various formats. Simplifying includes converting complex data into clear visualisations, media, or plain language summaries. This can include creating graphs, infographics, videos, memes, posters, postcards, or even poems. In her PhD thesis, Sarah Williams (2023) highlighted the enhanced impact that activist messages can have when delivered as hip-hop rap or spoken word poetry. Tailoring information involves crafting messages and incorporating images that align with the target audiences' values, beliefs, and interests to drive engagement with the subject matter and encourage them to share it within their own networks.

Capturing key messages in easily digestible and engaging formats helps amplify the information and promote memory retention. Additionally, creating an artifact that people connect to, and that stirs some emotional resonance, increases the chance that they will share it among their networks and those people will share it and so on. Translating vital evidence into bite-sized packages means it has greater potential to spread virally among the general public.

Sharing for advocacy

Sharing findings from evaluative activities is a form of advocacy. Real-world exemplars can magnify activist impact. By openly sharing their work, findings, and insights, everyday activists can contribute to an expanding repository of evidence where shared wisdom accelerates progress. Transparently communicating findings not only elevates credibility but also inspires and mobilises others. It raises the profile of the cause and helps the activists involved become known in the sector. At the rabbit sanctuary, Tanya noted that this resulted in other large organisations coming to them for advice because "despite [our] small team, [we've got our] name out there very effectively". Daisy explained that the standards upheld and the track record they demonstrate means that her organisation in Papua New Guinea is recognised by international governments as legitimate and respected. Alyssa and Ricki highlighted that being known meant that they are often asked to sit on panels and provide expert opinion in the media and to commissions. This section highlights some of the dissemination strategies used by the everyday activists interviewed for this book with a brief outline of the translated products that could be suitable for other activists.

Social media

Many of the interviewees are highly skilled at translating data into engaging and understandable products. The many social media platforms, such as TikTok, Facebook, Twitter/X, Instagram, Threads, and LinkedIn, provide an effective means to share bite-sized updates, data snapshots, and personal reflections. Visual content such as infographics, short videos, and compelling images can captivate attention and convey key insights. Alyssa creates memes about alarming declines in kangaroo numbers. She shares these on social media using photographs she and others have taken and data she has collected from secondary sources. Clariana uses Instagram to highlight body positive information using eye-catching graphs and other visualisations. Ricki uses information to weave video stories about inclusion and social justice for transgender people using platforms such as YouTube. As some of Alex's activism work involved her recycling efforts within a large organisation, she utilised the company's internal social media platform to display her activist achievements, which resulted in people at other branches asking for the same bins; thus, extending her impact.

Blogs

Blogs can be a useful method of providing new information and keeping followers updated on important facts and lessons. Kim runs a blog imbued with facts about bird protection and wildlife photography ethics alongside stunning images. Tamsin and Alyssa run a blog on green burials. Other interviewees, including Khalil, write for institutional blogs linked to think tanks, professional associations, and scholarly topics of interest. Everyday activists can write about their experiences, challenges, and successes, and weave in the monitoring data to substantiate the narrative. Alternatively, blogs can utilise monitoring and reflective data to write an advocacy piece aimed at garnering public support or influencing policy.

Newsletters

Dia's team compile their data and reflections into digestible formats for monthly email newsletters. These newsletters are utilised to update over 100 subscribers on progress, lessons learned, and future plans. Similarly, Caroline sent monthly emails that reported quantitative monitoring data on their recycling efforts. This can be an effective means of sharing information depending on the activist cause. The activist endeavours of our interviewees ranged from people who were part of an organised movement to those who adopt an ad-hoc approach, moving between several small-scale informal activities. Newsletters may be appropriate for the former, but probably not the latter.

Group work

Webinars and workshops can provide people with a deeper level of information than is possible through an infographic or a meme. These events are opportunities to present monitoring data, delve into reflection processes, and engage with participants through question-and-answer sessions. While they are a valuable technique, they are time intensive and may involve other resourcing needs. A benefit of holding these sessions online is that resource costs are reduced and the event can be recorded and posted elsewhere on websites and across social media to extend engagement. Further, these events need not be in a formal webinar or workshop format, they can be simple community gatherings or storytelling sessions, held online or in person, where monitoring data and reflections are presented and attendees are engaged in discussions or interactive activities.

Audio-visual productions

Recording podcasts or audio-visual segments offers another way for people to engage with evidence and humanises content and stories through sharing activists' authentic voice. Community radio stations are often willing to discuss pressing concerns, especially when they are locally relevant. Domuto can track the impact of his radio appearances as this dissemination method allows people in villages throughout the region to hear about his sports-focused peacebuilding activities. "We even get requests from youth in rebel held areas who are also in need of sports activities" he exclaimed. Alyssa discussed how she has built her media networks over time and now knows where to go with what information to amplify the chance of the story making it into the newspaper or being invited to attend media interviews. She explained that:

> Every media article we get I see that as a bit of a success. Every time we get quoted, I know there will be eyeballs on that. It's good to have those conversations with people, and maybe that discussion will get them thinking about things that they haven't thought about before.

Networking

Peer networks, both in person and virtual, offer a place to share information and insights to a knowledgeable and receptive crowd who can add layers to the evidence with their own critical reflection and feedback based on experiential wisdom. Speaking with other activists provides opportunities for two-way knowledge transfer and mentoring. Further, people who care about the same causes are more likely than the general public to follow and promote information that taps into shared passions and concerns. For example, Marilee speaks at conferences, forums, and with communities of practice about climate change. Attendees then share her takeaway messages on social media and repost her summaries of the event.

Reach can be extended through partnering with social media influencers or public figures who align with the activist cause. They can help amplify the message to a broader audience and mobilise new supporters. Ben highlighted the buzz of "getting in the ear of people with influence and then seeing those people talk about the issue you raised with them". In 2016 Marilee was staffing a booth at a climate change conference and engaged in a conversation with an evidently powerful man who was trailed by two bodyguards. The man was sceptical about climate change but shortly after the conference Marilee discovered that he was a high-ranking minister in a Middle Eastern nation. In the following year the country released its first climate change adaptation plan. While not attributing the plan to her conversation, she feels that her efforts may have contributed.

Informal or formal research

Collaborating with like-minded individuals, organisations, or activists to co-author articles, reports, or case studies that spotlight monitoring data and other evidence can extend the reach of everyday activists into different circles. Connecting with academics and think tank institutions may help provide credibility and promote policy briefs that government is more likely to notice than if it was written by a lone community member. Additionally, co-authoring with these groups makes the information more accessible to other academics and think tank institutions who may incorporate the findings into their own analyses and advocacy work and thereby increase impact. In terms of self-growth, Vijeta reflected on an academic paper she wrote with one of the authors of this book to highlight disenfranchised grief among culturally and linguistically diverse lesbians (Uppal & Kelly, 2020):

> Writing that paper with you really helped me strengthen my relationship with myself. And I presented my story on Power at a storytelling woman-only group in Melbourne in 2019. It was the first time I came out openly to a room full of strangers and something shifted inside me. I began to embrace my story and built a stronger connection with myself that helped to positively influence the world around me.

By combining these dissemination strategies, everyday activists can promote their monitoring data, experiences, and reflections to a wider audience, sparking meaningful conversations and inspiring positive change.

In practice

For people who want to explore evaluation in more depth, drawing upon established evaluation models and frameworks that are readily available on websites and in the literature may be a beneficial starting point. For example, frameworks such as Kirkpatrick's (1984) model for evaluating training programs could be useful when a greater degree of rigour is necessary (Kirkpatrick Partners, 2023). Using a collective impact approach to measuring collaborative efforts can help groups joining forces to paint a bigger picture (FSG, 2023). Results-based accountability can help home in on assessment of project results by asking: How much did we do? How well did we do it? Is anybody better off? (Clear Impact, 2016).

There are hundreds of evaluation approaches, any of which can help activists to think beyond just counting how many things are done. They help work towards articulating quality and impact, so that activists can determine whether the efforts are making a difference. However, tracking, reflecting, and sharing activist work can be as simple or as complex as fits each situation. It is essential to find a method or approach that provides relevant data, is appropriate for the project, and aligns with available resources.

The strategies discussed throughout this chapter may complement and extend the impact of activist efforts, as recapped here:

- Gather information to determine whether your approach is based in evidence.
- Develop a simple plan for tracking activist work.
- Consider whether you want to draw upon an existing approach or model to develop key questions.
- Collect and store relevant data to measure progress towards outcomes.
- Seek feedback through engaging with peers, mentors, or community members to gather external perspectives.
- Carve out space and time for reflective practice. Maintain a reflective journal or digital record to capture insights, challenges, and evolving perspectives.
- Simplify data by producing clear visualisations, infographics, or plain language summaries.
- Craft narratives and stories that highlight your real-world impact.
- Tailor messages to resonate with your audience's values, beliefs, and interests.
- Involve community members, recipients, partners, and other stakeholders in sharing your impact.
- Leverage diverse platforms such as social media, workshops, and community events to disseminate information.

Conclusion

Engaging with everyday evaluation can help you think deeply about your positionality. It can assess not only whether you are doing things right, but also whether you are doing the right things. Evaluation is not just a formality. It is a powerful tool to check appropriateness of responses, elevate activities, generate knowledge, inspire followers, and drive change. Evaluation can interrogate the very foundations of your approach. By embedding robust data practices, carving out moments for introspection, and welcoming external feedback, activists not only maximise their influence and awareness but become beacons of knowledge and inspiration for others.

While many activists initially claimed they do not evaluate, our discussions unveiled their innate evaluative actions, from tracking donations to recounting campaign tales. By demystifying monitoring and evaluation, this chapter has crafted a streamlined approach for everyday activists to amplify their impact and help them determine the best approach. Many of our interviewees have transformed their findings into vibrant memes, compelling videos, and impactful graphics. They take their messages far and wide through platforms like social media, community events, and policy briefs. But it is not just about broadcasting – it is also about refining. Integrating evaluation into your activist agenda can provide fresh insights that pave the way for adaptation, improvement, and greater positive impact.

References

Clear Impact. (2016). Results-based accountability. https://clearimpact.com/results-based-accountability/

Fetterman, D. (2003). Fetterman-House: A process use distinction and a theory. *New Directions for Evaluation, 2003*(97), 47–52. doi: https://doi.org/10.1002/ev.74

Fetterman, D. (2023). *Empowerment evaluation and social justice: Confronting the culture of silence.* Guilford Press.

FSG. (2023). *Collective impact forum.* An Initiative of FSG and the Aspen Institute Forum for Community Solution. https://collectiveimpactforum.org/

Gibbs, G. (1988). *Learning by doing: A guide to teaching and learning methods.* Further Education Unit.

Global Evaluation Initiative. (2023). *Better Evaluation.* https://www.betterevaluation.org/

House, E. (1990). Methodology and justice. In K. A. Sirotnik (Ed.), *Evaluation and social justice: Issues in public education. New Directions for Program Evaluation. No.* 45 (pp. 23–36). Jossey-Bass.

House, E. (2007). *Values in evaluation and social research*. SAGE Publications. doi: https://doi.org/10.4135/9781452243252.n1

Kelly, L. (2019). *What's the point? Program evaluation in small community development NGOs*. Doctoral Dissertation, Deakin University.

Kelly, L. (2021). *Evaluation in small development non-profits: Deadends, victories, and alternative routes*. Palgrave Macmillan. doi: https://doi.org/10.1007/978-3-030-58979-0

Kelly, L. (2022). Worthwhile or wasteful? Assessing the need for a radical revision of evaluation in small-sized development NGOs, *Development in Practice*, 32(2), 201–211. doi: https://doi.org/10.1080/09614524.2021.1937540

Kelly, L. & Htwe, P.P.T. (2023). Decolonizing community development evaluation in Rakhine State, Myanmar, *American Journal of Evaluation*, (online first), doi: https://doi.org/10.1177/10982140221146140

Kelly, L., & Reid, C. (2020). Baselines and monitoring: More than a means to measure the end. *Evaluation Journal of Australasia*, 21(1), 40–53. doi: https://doi.org/10.1177/1035719X20977522

Kelly, L., & Rogers, A. (2022). *Internal evaluation in non-profit organisations: Practitioner perspectives on theory, research, and practice*. Routledge. doi: https://doi.org/10.4324/9781003183006

Kirkpatrick, D. (1984). *Evaluating training programs: The four levels*. Berrett-Koehler Publishers.

Kirkpatrick Partners. (2023). The Kirkpatrick model. https://www.kirkpatrickpartners.com/the-kirkpatrick-model/

Markiewicz, A., & Patrick, I. (2016). *Developing monitoring and evaluation frameworks*. Sage Publications.

Patton, M. (2018). A historical perspective on the evolution of evaluative thinking. *New Directions for Evaluation, 2018*(158), 11–28. doi: https://doi.org/10.1002/ev.20325

Rogers, A., & Malla, C. (2019). Knowledge translation to enhance evaluation use: A case example. *The Foundation Review*, 11(1), 49–61. doi: https://doi.org/10.9707/1944-5660.1453

Saad, L. (2020). *Me and white supremacy: How to recognise your privilege, combat racism and change the world*. Quercus.

Schon, D. (1983). *The reflective practitioner: How professionals think in action*. Basic Books.

Uppal, V., & Kelly, L. (2020). Living on the margins: A South Asian migrant's experience of disenfranchized grief as an ethnic and sexual minority. *Journal of Gay & Lesbian Social Services*, 32(4), 502–516. doi: https://doi.org/10.1080/10538720.2020.1799284

Wadsworth, Y. (2011). *Everyday evaluation on the run*. Left Coast Press.

Western Michigan University. (2023). The evaluation center – Evaluation checklists. https://wmich.edu/evaluation/checklists

Williams, S. (2023). *Born to stand out: The role of hip hop for young South Sudanese Australians in building their political voice to resist racialising discourses*. PhD Thesis, Deakin University.

8 Self-care

A recurrent theme throughout the interviews with everyday activists was the consequences of neglecting self-care. Many began their narratives by recounting the impacts of burnout. This is an all-too-familiar outcome of prolonged exposure to chronic stress, particularly in the demanding and socially impactful contexts that define activism. As activists pour their energy into addressing societal challenges, the weight of stress accumulates, heightening the risk of burnout. Even those who had not reached the point of complete burnout could resonate with feelings of being overwhelmed, overworked, and fatigued – resulting in emotional, mental, and physical consequences. Many interviewees shared examples of having to step back from jobs, relinquish volunteer roles, or distance themselves from challenging situations. These instances underscored how burnout and exhaustion can hinder effectiveness and compel individuals to deviate from their original intentions.

Activists who neglect self-care may find themselves drained and less capable of engaging empathetically with others. It can foster a negative shift in attitude, resulting in cynicism and depersonalisation. Activists may develop a detached and callous stance towards their cause, their colleagues, and even the very individuals whom they seek to support. They may perceive their efforts as ineffectual or futile, eroding their motivation and causing them to question the value of their work. These changes in thinking undermine an activist's capacity to contribute optimally to their cause.

Excessive dedication to activism at the expense of personal relationships can cause activists to become socially isolated. Neglecting meaningful connections with others can create a ripple effect, compromising support networks and exacerbating feelings of isolation. Genuine communication can break down and it becomes even more difficult to advocate using genuine compassion. This means the activist's message is less likely to resonate and hampers their capacity to evoke meaningful change.

DOI: 10.4324/9781003333982-8

Ultimately, working excessively without an adequate self-care regimen can precipitate waning enthusiasm and an overall loss of motivation.

Burnout (and the road toward burnout) is a serious mental health problem. Burnout manifests as emotional, physical, and cognitive exhaustion, accompanied by detachment, reduced empathetic engagement, lack of confidence, and diminished personal accomplishment (Smith, 2022). Burnout is not sudden in onset; it occurs when stress responses are repeatedly triggered over the long term without sufficient opportunities to rest or restore. Burnout evolves gradually over time, often fuelled by a combination of demanding workloads, unrealistic expectations, and a lack of effective coping mechanisms. Clinical psychologist Dr Julia Smith (2022) in her book *Why has nobody told me this before?* wrote that there is often a mismatch between the person with burnout and the level of control they have over their resources, the amount of recognition or acknowledgment they receive, or the level of social support they can access. The person may also perceive a degree of unfairness or inequity, or find that the tasks they are undertaking conflict with their values. As activists focus their energy on addressing societal challenges, the cumulative toll of unrelenting stressors erodes their wellbeing. Persistently toiling without adequate intervals for respite can precipitate physical weariness, disturbed sleep, irritability, lack of concentration, compromised immune functionality, and mental fatigue. A person can feel overwhelmed and find small problems unmanageable. This state renders activists susceptible to both physiological ailments and psychological distress (Smith, 2022).

Therefore, it is important to recognise, address, and mitigate burnout. Acknowledging burnout as a legitimate threat is the first step towards adopting comprehensive and proactive self-care measures. Self-care encompasses a spectrum of intentional practices that prioritise physical, emotional, and mental wellbeing. By integrating self-care practices into routines, activists can enhance their resilience, bolster their emotional equilibrium, and fortify their cognitive capacities. Empowering activists with the tools and knowledge to safeguard their wellbeing is essential for nurturing their sustained commitment and cultivating a more compassionate and effective activist community. Purposefully incorporating and prioritising self-care practices bolsters the resilience of everyday activists, thereby preserving their effectiveness and passion in advancing their causes.

Interviewees offered an array of strategies for promoting and safeguarding wellbeing because they recognised the crucial significance of self-care, not only for their personal wellbeing but also for their families, friends, and causes. These strategies are detailed throughout this chapter, encompassing elements such as meeting basic needs, reframing challenges, aligning actions with strengths, managing demands and resources, nurturing autonomy, practicing mindfulness and self-compassion, and upholding boundaries. In the following pages, we delve into the heart of self-care in activism, weaving together the wisdom of both psychological insights and the voices of those who are in the thick of it. Through their experiences, strategies, and insights, we highlight that the preservation of one's own wellbeing is not just a necessity, but a fundamental requirement for enduring transformation.

Strategies for self-care

The psychological theories on self-care will be interwoven with data gleaned from the interviews with everyday activists. These theories include the hierarchy of needs, cognitive appraisal theory, strengths-based approaches, the job demands-resources model, self-determination theory, mindfulness and self-compassion, and boundary management theory. By understanding and embracing these theories, activists can cultivate proactive strategies that empower them to contribute more effectively, engage more compassionately, and thrive amidst the challenges of their activism.

Ensure needs are met

Abraham Maslow's (1943) hierarchy of needs contends that individuals need to fulfil basic physiological needs, such as food, rest, and safety, before moving up the hierarchy to meet higher-level needs like belonging, self-esteem, and self-actualisation. Thus, ensuring adequate rest and healthy nutrition provides a solid platform to effectively pursue goals.

Underscoring the symbiotic relationship between activism and self-care, Liz and Domuto spoke of how they prioritise their health through exercise and nutrition to sustain their contributions. Similarly, Mike and Adrian talked about eating nutritious foods and taking care of their bodies and minds. This included finding things to be grateful for and laughing as often as possible. Domuto, Ronny, and Alyssa highlighted the importance of getting enough sleep. Vijeta's integration of wellbeing into her ethics underscores how nourishing herself translates into enhanced social impact: "For me the journey starts with my own self, if I'm a happy soul then I'm having less of a negative impact on society. So I am always working on myself." Vijeta has integrated self-care into her life and routine.

When asked what Tamsin does to look after herself, she responded with "whiskey". While this may sound a little flippant at first, she does not mean that she drinks herself into oblivion to cope with the demands and pressures of balancing activism work, her own small business, and four young children. This is a structured technique to ensure she has some special downtime each day. She said:

> Between 4 and 5pm I like to sit here on my kitchen bench with a book and a whiskey and not really engage with anything else. The kids know that for that hour I'm pretty much not available. Some days I really need that hour, and sometimes not so much. I really try to adhere to it as well. It's important. It's nice, I like it.

Similarly, Clariana highlighted the value of time out: "I take a Me Day, not doing any work, but just doing all the things that make me calm and bring me back to me. I try and remind myself of many wins, and the impact that has on people." Nicole books two holidays a year so that she always has something to look forward to and has an end in sight. Tamsin's cherished ritual, Clariana's Me Day, and Nicole's annual plans resonate with the imperative to prioritise oneself. Recognising the need for rest, relaxation, and self-compassion, these activists rejuvenate their physical and emotional wellbeing.

Many of our interviewees discussed the potency of human connection in self-care. For example, Marion recognised that, "social connections are really important for me. What really helps to sustain me I think is the friendship and the support of others." Interviewees noted that engaging with family and cultural networks provides emotional sustenance, reaffirming one's identity and values. As an Aboriginal Australian, Jody's journey to Country resonates with the same essence, highlighting the nourishment derived from reconnecting with ancestral land, sea, sky, and spirit.

Everyday activists draw from diverse sources to sustain their passions. Iram noted, "I spend quality time with my family, going out meeting friends, watching a movie, or a beach walk. I understand that social activists like us need to unwind themselves in order to do better. I like to cook and share food with my friends and family." Susanne commented that for time out she would, "walk the dog, go to the beach, pat the cat". Ricki also highlighted the comfort of patting her cats. Many mentioned music, with Ronny explaining that he likes to "just pick up the guitar and play music" while youngsters Talia and Clariana have found crochet and other crafts to be a calming retreat. Ronny's music, Talia's creative pursuits, and Ricki's affinity for cats exemplify the array of avenues through which they replenish their reserves. The variety of methods reflect the unique essence of each activist's self-care journey.

Without self-care, activism drains power and depletes. This sentiment was reinforced by Dia who noted: "Self-care is absolutely paramount... You can't be efficient and effective without looking after yourself." Marilee emphasised: "I'm judicious about applying self-care practices; about taking rest and doing things that give me joy." Vijeta offered that it is important to enjoy being alive by "doing something fun and playful". She talked about her dog and how he brings a lot of playfulness into her life. Phyo Phae Thida explained: "Listen to your body. Sometimes you're just so tired and you can't concentrate. Your body is physically telling you to take a break and do something else." Self-care weaves harmony into an activist's life. Looking after yourself helps you better look after others. Through striving to strike a balance between activism and self-nourishment, these activists demonstrate the interplay between giving and replenishing.

Reframe challenges

Stressors will always exist within a perpetual cycle. Yet, within this cycle lies the potential for growth – a realisation that each trial serves as a stepping stone on the path to greater understanding and action. Recognising this phenomenon, cognitive appraisal theory by psychologist Richard Lazarus illuminates the powerful role of perception in shaping how activists navigate stressors. When stressors can be viewed as opportunities for growth and learning, activists can enhance their emotional resilience because their efforts are understood to be steps in a long-term transformative journey (Lazarus & Folkman, 1984; Manson, 2017).

Reframing is a powerful cognitive tool, enabling activists to reclaim their narrative from negativity. Dia asserts that toxic relationships, whether familial or social, can be managed by shifting one's perspective. Through reframing, activists protect their mental wellbeing while channelling energy into more productive pursuits. For example, instead of becoming frustrated when engaging in discussions with a loved one who is cognitively inflexible, reframing this as an opportunity to hone your arguments and consolidate your thoughts can prevent feeling like you are wasting your breath.

Embodying the heart of cognitive appraisal theory, Mike urges activists to accept the paradoxes of existence – a world both perfect and fraught with suffering. He said, "When we can hold that paradox in our minds and walk that tightrope without spinning out, then we can negotiate a positive path through life." He explained that he makes sure to:

> ... remember every day that I'm not somehow magically separate from it like an ego walking around in a bag of skin and bones.

Find something of beauty and awe in every single day. Try to see the great rolling drama of life and existence as a dance and a game, rather than a tragedy and a disaster.

Activists, united by shared frustrations, confront the reality that not all causes resonate equally. Kim commented that:

> Something I come across quite often is people getting really upset when they're really passionate about something and there's not enough money to fight it or enough people caring about that particular thing. And something I find myself saying is "we all feel that". It's not just that particular thing you care about... it might be the Lord Howe Island stick insect, but whatever it is that we're passionate about, not everyone is going to be that passionate about it. People need to realise that we are all passionate about different things. There will be people who oppose it and people who are totally apathetic to it. I know some people find that very frustrating. It is disappointing, but it's not just their thing that people don't care about, it's all of our things. We all feel that way. I mean, if every decent person across Australia said: "We need to get those refugees out of detention," I feel like if enough of us did that, they'd be out. And that's heartbreaking. I think that's the same with all of these things, we just have to keep being drops of water on a stone, and looking after ourselves and not losing hope. [This includes] realising that different people will put varying amounts of effort into different causes and being at peace with that.

Kim's call to "keep being drops of water on a stone" encapsulates the essence of persistence. By reframing the collective impact, activists derive strength from their joint efforts, regardless of individual outcomes.

Ricki, Matt, and Nigel highlighted the value of optimistic thinking and searching for the good in every situation, including traumatic personal experiences. Marilee takes this a step further by journalling five good things that happened each day. Ben reframes things by undertaking "professional development and trying to learn different ways of doing things, different techniques". Adrian noted the value of reconnecting with the facts of his activist topic to try and put his work into perspective. Similarly, Alicia explained how she reconnects with the people she seeks to serve at the community level to refresh perspective on her life:

> I actually travel out to these women and places where I feel like I'm home. I go to villages, because that's how I grew up, and it just

gives me perspective when I see their life, and I appreciate what I have, and then I come back re-energised, determined to help more.

As well as spending time on self-reflection, Thiha and Phyo Phae Thida mentioned that they appreciate having opportunities to engage in reflective practice with others working in a similar space. This was identified as a way of stepping back from the work, seeing it from other perspectives, and working towards solutions that reframe intractable problems into manageable tasks.

Vijeta and Nicole spoke of channelling their anger against racism and lateral violence to motivate their activist pursuits; thus, illuminating a path for activists to utilise their emotional turmoil as a driving force. By transforming anguish into agency, activists gain a newfound sense of purpose, lending their energy to meaningful change.

Cognitive appraisal theory encourages activists to reframe stressors into something helpful. Letterbox drops need not be overwhelming if they can be considered as an opportunity for a relaxing walk. Instead of feeling rushed about getting to a demonstration on time, it may be a chance to catch up on an interesting podcast. Organising a major event or managing a team with challenging interpersonal dynamics can build transferrable skills. Experiencing something awful can provide insight, empathy, and post-traumatic growth.

Utilise personal strengths

Identifying and leveraging personal strengths can enhance resilience and motivation. Embracing a strengths-based perspective enables activists to recognise their unique talents and capacities (Clifton & Harter, 2003; Seligman, 2002). By aligning their efforts with their strengths, activists maximise their effectiveness and derive intrinsic satisfaction, fostering a sustainable commitment to their cause (Buckingham, 2008). Drawing from positive psychology, the act of identifying and embracing one's strengths is a useful tool for cultivating resilience and maintaining motivation (Clifton & Harter, 2003; Seligman, 2002).

A strengths-based perspective can offer a transformative lens. Where each person is a piece of the puzzle, activists contribute their various strengths to create a vibrant and impactful whole. Jenny, with her love for people and careful planning, flourishes as an event host where she can focus on executing the occasion. Alex, who finds solace in community gardening, sees her activism as an extension of self-care because it is something she does well and enjoys. Nigel was feeling overwhelmed by the acceleration of technology in recent years and realised that his area of activism was not working to his strengths and was draining his energy. As he pivoted towards his passion for planning and the environment, he discovered a rekindled sense of fulfilment. When one's capabilities merge with their purpose, it can create a synergy that fuels effective and sustainable activism.

Effective activism requires many skills beyond the clichéd role of standing up in a protest with a loudspeaker. Tanya, recognising her aversion to crowds and confrontation, thrives as part of the support team where she skilfully conducts the essential work on which organisational survival depends. She candidly speaks of her social anxiety, acknowledging that some forms of activism are beyond her comfort zone. Her resilience emerges from the understanding that each activist, regardless of their individual struggles, has a unique contribution to make. Tanya identifies personal limitations and reframes them as triumphs:

> I have to accept that I have social anxieties. I can't go out and do things I so admire in other people. For example, those who go out and take undercover footage, they're just heroes to me. There's no way I could ever consider doing something like that… When I finally accepted in myself that I just don't have the capacity to do that and instead focus on the things I can do, I became less overwhelmed and more effective. I still feel guilty sometimes that I can't do those things. But I put all of my effort into the things that I can do.

This self-awareness steers her towards tasks where her strengths can shine. Resonating with Clifton and Harter's (2003) principles, Tanya advocates for introverts and people with social anxieties to channel their talents into areas such as administrative support, letter writing, online petitioning, creating memes, and crafting social media posts from the comfort of their own spaces. Tanya's argument also highlights the need for everyday activists to acknowledge quiet contributions and not simply reward those with the loudest voices and most flamboyant displays. Tanya appreciates that her peers value her contributions and note that she possesses skills that they do not. This underscores the reinforcing loop between strengths, motivation, and the sense of meaningful contribution.

Assess resource needs

Arnold B. Bakker's model champions the art of equilibrium – the balancing act between the demands activists confront and the resources at their disposal (Bakker & Demerouti, 2007). His *job demands-resources model* focuses on the balance between demands required to do the job and available resources. This approach could help activists identify stressors related to their cause and proactively seek support or resources to manage demands effectively. By aligning demands with available resources, activists optimise their work environment and reduce the likelihood of burnout.

Bakker's model prompts activists to break down colossal challenges into manageable components. Dia underscores the importance of simplification and Alex identifies the value of "helping yourself step through the steps of how we're going to get there". This approach empowers activists to discern stress-inducing aspects and identify areas where additional resources or support could alleviate the strain.

Elise's experience volunteering within a group that encourages people to divest from traditional financial institutions and invest their money in ethically sustainable ventures beautifully aligns with Bakker's principles. By offering a flexible engagement scale, the group empowers activists to evaluate their capacity and modulate involvement according to external commitments and capacity. The model's wisdom lies in recognising that an activist's journey is multi-dimensional, and harmonising involvement with personal circumstances is the key. This approach helped Elise maintain engagement with the group. Rather than feeling stressed or pressured to do more, she knew she was appreciated and could return when the time was right for her.

The model emphasises the often-neglected power of delegation. Matt's insights resonate with Bakker's teachings – delegation acts as a form of self-care. Khalil advises that sharing responsibilities prevents burnout, cultivates teamwork, and offers different ways of tackling problems based on people's varying skillsets and experience. Kim offers the people she works with diverse avenues for engagement, so they can optimally use their resources and foster sustainable growth for both the cause and the individuals involved. Dia involves the next generation of everyday activists by identifying meaningful tasks for young people: "We have junior angels too where younger people come and do little things like packing hampers and we hope that one day this legacy will be taken over by our children. Now they are bringing their own skill set, like someone is into music and some into art."

Bakker's model not only considers the available resources but also advocates strengthening collaboration and support networks. Dia envisions that activists forge networks that extend beyond work, becoming families bound by shared aspirations. This sense of community rejuvenates spirits and fans the flames of activism. Dia's model goes a step further with her intentional buddy system whereby activists are carefully paired with someone else in the organisation who looks out for them, checks in, and supports their wellbeing. Everyday activists who work in ways that align with Bakker's principles often attract likeminded individuals or discover resources that enrich their journey. This synchronicity, a manifestation of the law of attraction, reiterates that aligning intent and focus with available resources paves a smoother path to change. Within the context of everyday activism, Bakker's *job demands-resources model* offers a philosophy of equilibrium, empowerment, collaboration, and resilience (Bakker & Demerouti, 2007). Susanne learned from previous experience that, "When you're constantly putting ideas forward and nothing is happening as a result, that can be really demotivating." Recognising that she was "casting pearls to swine", she quit before she became exhausted and went to another organisation that harnessed her ideas and motivation to enact change.

Maintain self-determination

Psychologists Edward Deci and Richard Ryan expanded self-determination theory to unpack the connections between autonomy, intrinsic motivation, and psychological wellbeing (Deci & Ryan, 1985). According to this theory, activists who cultivate a sense of autonomy in their advocacy efforts, continuously develop their skills, and foster meaningful connections within their cause are more likely to experience sustained wellbeing. This theory emphasises the importance of self-directed pursuits and individual agency around how and when activists are involved.

Of the everyday activists we interviewed for this book, Dia in particular emphasised a strong focus on self-determination. She believes it is important to be the boss of your own life and take responsibility for your actions and inactions:

> Whatever you thought you sacrificed… is undone if you don't look after yourself. Self-independence, self-reliance, and self-respect are core for going through life's journey. Don't blame others for an empty you. Don't say "I took care of you so now I am like this." Take responsibility for yourself. Life is long. Technically it's 80 years. The phases of sacrifice happen sometimes, but you cannot make it your personality.

This autonomy acts as a beacon, reminding activists that their path is shaped not solely by external demands but also by the choices they make and the boundaries they establish.

Embedded within self-determination theory is the notion of intrinsic motivation – the internally led driving force that propels activists forward on their path. As activists become attuned to their intrinsic passions, inspiration fuels their actions. Dia's words resonate deeply, urging activists to recognise their innate power and potential:

> It's the value creation, your legacy. What did you do today? For you or for what you love doing? Keep adding to that jar of value creation. Then that's a ripple effect that impacts your home, and your community. Every day wake up with focus, and it has to be *your* focus, and that is part of your self love.

Within self-determination theory, nurturing connections within the activist community is noted as significant. The interplay between autonomy, intrinsic motivation, and psychological wellbeing finds harmony in the bonds activists forge with like-minded souls. Dia explained how the activist women she collaborates with empower themselves through self-care, self-love, and the art of setting boundaries. This activist community is a place of sustenance, where individuals stand strong individually and collectively. Cultivating autonomy, intrinsic motivation, and meaningful connections paves the path to sustained wellbeing. Practicing self-direction and self-love is not an indulgence; it is an investment in oneself that resonates far beyond, influencing the lives of others and fostering a positive legacy.

Exercise mindfulness and self-compassion

Mindfulness practices can help activists stay present and manage stress. Self-compassion encourages treating oneself with kindness during challenging times, reducing self-criticism and perfectionism. Integrating mindfulness practices allows activists to cultivate present-moment awareness and manage stressors effectively. Simultaneously, self-compassion counteracts harsh self-judgment and fosters emotional resilience and self-kindness (Kabat-Zinn, 1990; Neff, 2003). Just as Australian author Anne Deveson (2004) aptly pointed out in her discussion with Geraldine Doogue on the current affairs program *Compass*, being resilient sometimes means allowing oneself to be cared for by others. An interesting observation by Professor of business Scott Galloway (2019) in his book *The algebra of happiness* is that we pull out of platonic hugs within a few seconds. The small act of soaking in a hug from a loved one is a powerful and simple way of allowing ourselves to be cared for by others.

Embedded within everyday activists' narratives are comments that resonate with the essence of self-compassion and mindfulness. Vijeta highlights the significance of embracing our own limitations: "I'm very mindful that I'm not going to save the world. Sometimes you have to create that boundary and be realistic about your possible impact." Caroline's reflection on nature's serene pace instils a sense of solace:

> I often think about something that a friend told me many many years ago when we were walking in Spain. He was looking at some stones and [said]: "Those stones, they don't care at all about the world." ... I think it's so peaceful and so true. Nature just does its business. It doesn't care about those things and just brings so much support and so much goodness. It really gives me some peace and energy.

Alicia explains how her self-care centres on protecting her mental health. "I do things that help me with my mental health, because I have a tendency, even though I tell people not to lose hope, ... I look at myself as a failure. So, from that point, I have to really focus on my mental health." Alicia's candid reflection mirrors the importance of prioritising mental health, acknowledging that even the strongest people must attend to their own hearts to sustain their transformative spirit.

Immersion in nature was regularly mentioned as replenishing. Khalil, Adrian, and Kim immerse themselves in nature, feeling at one with the trees around them and the ground beneath their feet. Khalil explains: "I go and see the trees and intentionally touch the tree, that's magic to me, or just sitting and looking at the water." Adrian said, "I get immense satisfaction out of getting my hands into soil or potting mix and just that process of communing with nature whether it's here or down the creek." Many people, including Adrian, Alex, and Susanne, mention gardening. Clariana highlights this as a nurturing, grounding activity: "I love plants, I have so many plants, so I will water them, clean their leaves. Anything that grounds me." Further, Daisy's retreat to nature to explore rivers, Kim's patient birdwatching in the forest, and Caroline's quiet picnics in Litchfield National Park underscore the power of natural spaces in replenishing the spirit. These retreats serve as sanctuaries for activists to rejuvenate amidst the chaos of their activist work. Whether in solitary contemplation or social connection, such sanctuaries restore equilibrium.

Self-compassion and mindfulness are not mere practices but gifts activists can give themselves. Aruna outlines her strategies: "I do meditation, breathing exercises, listen to music, especially I love raindrop or waves music. I read a lot, reading is meditation for me as well and when I want to express myself, I write poems." Noraini literally buys herself presents in a beautiful act of self-love:

> I get fresh flowers and put them on my bedside. It's me gifting them to myself... I don't need someone to give me flowers, I can get them for myself because I deserve it. It's my wellbeing fund, and that's important... just know that you don't need other people to value you, I can value myself. I should give myself appreciation, that makes my day happier.

Explaining the reasons behind her acts of self-love, Noraini expands that:

> Caring for me is not selfish, it's so I can do this thing more and more and continue to do it, instead of just being on reserve battery. I try to imagine a cell phone, once you're on that red blinking thing, you're just going to collapse. And it will take so much longer to get fully charged again. Once your battery is fully drained, it takes forever to get back. But if it's not that depleted you have much more to work with and it's much easier to recharge... I need to make sure that I'm not too depleted before I plug myself back in to a recharging source – And that could be anything, a camping trip with friends, coffee and croissant, a really good gelato, or a lemon curd. Simple inexpensive stuff that you really enjoy. Just a cup of properly made almond milk flat white [coffee] and a crispy croissant would start my day strong. That's an investment to my wellbeing.

While Noraini spoke of treating herself, Marilee's self-care includes creating a physical sanctuary: "I'm really intentional about beauty in my space and surrounding myself with things that give me joy. So like art and plants. When we bought this place, I was very intentional about having green out of every window and seeing the sky." Other interviewees highlight the benefits of talking to professional therapists as this provides them with dedicated space and structure to unpack challenges, build clarity, and develop solutions.

Many of the incredible people interviewed for this book mentioned that they sometimes consider themselves to be failures. Liz recognised that she tries too hard to be perfect. Tanya sees that she is very hard on herself but notes that it is her who sets herself these unrealistic demands: "I'm the only one who expects that much from myself." She suggests that reinforcing self-compassion is something that activist groups could promote:

There needs to be a lot more focus from groups when they're trying to get people involved, this form of activism isn't for everyone and that's ok. We do need to be told it's ok, you're not a failure, you're not letting the animals down, and we can find something that will fit with your needs.

Marilee offers this support to her fellow activists: "Staying aware and doing everything you can within the boundaries of what's possible for you is about all you can do." Similarly, Kim highlights that:

It depends on where we're at in our lives and what else is happening. It's important to pace ourselves. Take a break and don't feel bad about it. There's a fabulous quote about that. Birds don't all sing at once. In a choir, the singers don't all hold the note consistently. We don't have to sustain our campaign at full energy for it to be successful. We do need to look after ourselves first.

Dia concurs, noting that:

When you're on a plane and the oxygen mask drops, you need to take care of yourself first. If you haven't looked after yourself, you can't look after others. Everyone has to compromise, but if you are consistently at the bottom of your list of priorities, you are kidding yourself.

Self-compassion and self-love are not selfish. They enable everyday activists to transcend pressures and pitfalls, acknowledging that imperfections and moments of struggle are as much a part of the journey as triumphs.

Set clear boundaries

The art of setting and implementing boundaries is a vital facet of nurturing wellbeing (Harper, 2020; Manson, 2017; Reid, 2012). Everyday activists should set clear boundaries between their activism and personal life, ensuring dedicated time for relaxation, hobbies, and relationships. By delineating clear boundaries between their advocacy efforts and personal life, activists preserve their wellbeing and prevent burnout, enabling sustained engagement. Boundaries as described by our interviewees include the importance of focusing narrowly on specific topics, taking on only as much as one can comfortably manage, distancing oneself from negativity, and balancing activism with sacrosanct time allocated for self-care.

Boundaries shield activists from the overwhelming barrage of information and issues that besiege their attention and support them to identify what is really important to them and accept that locus of control (Manson, 2017). Recognising the multifaceted nature of burnout, Kim highlights the significance of narrowing one's focus, acknowledging that the constant deluge of global turmoil can require a strategic retreat to local issues for effective coping. Kim outlines that, "We need to accept that we can't be fully active in all the areas we'd probably like to be. I find myself consciously thanking others for caring about issues in ways I can't at the moment." Alyssa echoes this sentiment, admitting that becoming immersed in too many causes can lead to obsession and an eventual breaking point: "I get so obsessed with certain causes that I think about them 24/7. Sooner or later, I think I'll probably snap and have to leave."

Acknowledging the potential of burnout, Alyssa advocates for safeguarding emotional equilibrium through conscious boundary setting around areas of activism:

> A major self-care thing I do is putting parameters around the causes I deal with. I care so much about stuff involving farmed animals, climate change, vivisection, logging, refugees, etcetera – so many things – but I actually look at these things very little. I find I just get overwhelmed and feel hopeless, so I deliberately don't go there and focus on my little sphere. I think if I tried to help with everything, I would actually do nothing.

Elise, Jenny, and Serena concur, articulating how deliberate boundaries allowed them to channel their energy effectively, narrowing their focus within their respective spheres of influence. This process of focus encapsulates the ethos of self-determination discussed earlier in this chapter, where activists align their efforts with their capabilities to preserve their passion and resilience. Further, narrowing the focus of activism supports effectiveness as activists can develop expertise in that area. They can strengthen resilience by not trying to fix the overwhelming landscape of global problems, as outlined in Chapter Two. While these boundaries of scope are important, they should be flexible and allow for manageable actions to support the work of fellow activists such as signing a petition or sharing a social media post.

Balancing activism involves navigating the fine line between involvement and overextension. Here, boundaries assume the role of discerning gatekeepers. Dia's testimony of how her boundaries and recognition of her limits have become stronger over time underscores the maturation of self-care practices with experience. The art of discernment is echoed by Matt, who describes the value of mindful commitment. Matt notes that so many ideas are exciting but committing only to those he deems manageable allows contribution without overburden. Serena reiterates this principle, advocating for quality over quantity: "One principle in my life is less is more." Within this theme, boundaries and strengths-based approaches converge, as activists leverage their strengths to allocate their limited resources judiciously.

Saying "no" was mentioned by several interviewees, who noted that they were at different stages of implementing this simple yet tricky strategy. Noraini explains that she works around this by providing people with options; perhaps leading a project is out of the question but providing advice is possible. "No" can be a useful tool to help craft the optimal workload for sustained commitment (Zahariades, 2021).

Boundaries were also mentioned in relation to dealing with difficult personalities. A few people interviewed for this book suggested sometimes, depending on the situation and the personalities involved, it was necessary to step away from a relationship with a fellow activist who may have burnt out, become destructive, or otherwise ineffectual. Distance was important for the interviewees to protect themselves from infectious negativity. Some shared some awful experiences where fellow activists were disruptive, cruel, and harmful. Highly charged negative interactions may be caused by underlying issues associated with the activism or caused by other personal factors, such as post-traumatic stress disorder. Our interviewees suggested that activists should immediately distance themselves from people if they become destructive or slanderous. They were concerned that energy could be lost in trying to *fix* a person or (re)establish a healthy connection. Interviewees reflected on where their strengths lie and where their energy is best placed before reacting. Mike recognised that engaging with these people poses a risk to his own health, "I find that I am able to empathise with their situation all too easily." Dia indicated that she actively avoids conflict with people around her causes: "I don't judge if people don't want to get involved. It's something I've freed myself from because it's a black hole."

There are plenty of examples among our interviewees of overwork and self-identified problems implementing and maintaining boundaries. However, each of the interviewees had at least one strategy that they could share, and a few have been able to execute inspirational boundary management plans. The criticality of carving out dedicated moments for rejuvenation is exemplified by everyday activists including Dia, Serena, and Delphine who allocate specific days for rest and self-care and have communicated these to their networks. Similarly, Vijeta, Marilee, and

Susanne note that they effectively prioritise and maintain their boundaries through living within structured routines that include ensuring healthy habits and strict separation of activism and personal spheres.

Embedding boundaries can be hard and takes practice and determination, especially when dealing with people who are intent on consistently pushing against and disrespecting boundaries. Guidance from books such as Dr Faith Harper's (2020) *Unf*ck your boundaries: Build better relationships through consent, communication, and expressing your needs* and Mark Manson's (2017) *The subtle art of not giving a f*ck: A counterintuitive approach to living a good life* offer helpful advice for rewiring our minds and standing true in our authentic selves.

Reflection

Drawing from the experiential insights provided by interviewees, the spectrum of self-care practices they promote align with theoretical underpinnings. Maslow's hierarchy of needs provides a roadmap for navigating personal growth (Maslow, 1943). By nurturing wellbeing, activists empower themselves to channel their passion into lasting change. Cognitive appraisal theory supports activists to rewrite the script of stress (Lazarus & Folkman, 1984). By reframing their perceptions, activists metamorphose challenges into catalysts for growth, fostering resilience, and propelling the wheels of change forward. Drawing upon strengths-based theory, activists can feel good about their contribution because they are drawing from their skill set instead of struggling to offer something they find hard and unenjoyable (Buckingham, 2008). Through combining activists with a diversity of skills and interests, each strength adds vibrancy and depth to the collective (Clifton & Harter, 2003; Seligman, 2002).

Bakker's model provides a practical guide to defend activists against burnout and amplify their contributions by ensuring that required tasks are met with suitable resources (Bakker & Demerouti, 2007). Guided by self-determination theory, activists can stay true to themselves and safeguard wellbeing by embracing autonomy, nurturing intrinsic motivation, and fostering connections (Deci & Ryan, 1985). By prioritising themselves and their interests, they are more likely to remain engaged, passionate, and active in their chosen space; an intrinsically motivated person who determines their own direction has a strong foundation for effective activism. The amalgamation of self-compassion and mindfulness practices equips everyday activists with a toolkit to navigate demands while nurturing their emotional wellbeing. By cultivating self-kindness, acknowledging the imperfections inherent in the journey, and grounding themselves in the present moment, activists can enhance their resilience, sustain their commitment, and effect positive change (Kabat-Zinn, 1990; Neff, 2003).

Finally, a common thread that is highlighted across all these themes is the importance of maintaining healthy boundaries (Harper, 2020; Manson, 2017; Reid, 2012). Rather than having a frantic experience of activism culminating in burnout and disengagement, everyday activists can enjoy making a difference at a time and place that suits them while ensuring plenty of downtime to recharge. As such, there are tangible benefits in delimiting the scope of focus, embracing a workload in alignment with one's capacity, and cherishing self-care. The following list of practical strategies may help everyday activists safeguard their wellbeing, sustain their advocacy efforts, and create a healthy balance between their personal and activist roles:

Ensure needs are met

- Make a commitment to look after yourself and prioritise personal health.
- Engage in activities that ignite joy and foster a sense of rejuvenation.
- Cultivate strong connections with friends, family, and mentors for emotional sustenance.
- Create a self-care routine that prioritises and takes control of your wellbeing. Incorporate a diverse mix of activities that uplift your spirit, ensuring a well-rounded approach to wellbeing.

Reframe challenges

- Develop a habit of reframing challenges into opportunities for growth. Consciously shift your perspective to see the potential lessons in adversity.
- Enhance self-awareness by documenting stressors and how they have contributed to your personal development.
- Nurture a mindset of resilience and growth by integrating gratitude into your daily routine.

Utilise personal strengths

- Tailor your activism to align to your unique skills, qualities, and traits.
- Team up with activists who possess complementary strengths to create a blend of talents.
- Enhance your strengths through courses, mentorship, or self-guided learning.

- Identify and embrace your limitations to find roles that accommodate your comfort zone, allowing you to flourish.

Assess available resources

- Identify the resources at your disposal – time, skills, assets, financial support, and networks – to aid efficient allocation and prevent resource overuse.
- Stocktake the demands of your activism efforts to gauge if demands and resources are balanced.
- Break down goals into smaller manageable tasks.
- Recognise gaps and seek resources or support through encouraging sharing with other activists.
- Collaboratively develop a delegation protocol, distributing responsibilities among a team.

Maintain self-determination

- Define your personal values and the causes that resonate with you.
- Set achievable goals that align with your passions, giving you a sense of control and ownership.
- Identify the aspects of activism that bring you joy and fulfilment. Focus on activities that align with your intrinsic passions. Embrace what truly motivates you to ensure that your efforts remain authentic and sustainable.
- Seek out like-minded individuals or groups who share your passions and values to create a network that fosters a sense of belonging and solidarity.

Exercise mindfulness and self-compassion

- Infuse mindfulness into your daily routine. Engage in tasks with focused attention to create opportunities for present-moment awareness that help manage stress and foster a sense of calm.
- Practice deep breathing exercises and meditation to anchor yourself in the present moment.
- Engage in kind and compassionate self-talk to counter self-criticism and perfectionism.
- Open up to trusted friends, family, fellow activists, or access professional debriefing and counselling services for a structured platform to process challenges and emotions.

Set clear boundaries

- Define specific working hours and spaces for activism-related tasks. Once those hours are over, switch off from work-related communications and activities to focus on personal time.

- Openly communicate your boundaries with colleagues, family, and friends. Establish clear guidelines for communication channels and response times.
- Learn to say no. Practice assertiveness by declining additional commitments without guilt, emphasising the need to maintain a balanced workload.
- Raise awareness about the significance of boundary management within your activist community.
- Encourage others to engage in respectful boundary-setting to foster a culture of wellbeing.

Conclusion

This chapter examined the relationship between psychological theories and practical self-care strategies, offering a roadmap for sustainable activism toward resilience and health. The chapter aimed to empower everyday activists to accept the profound importance of self-care. Self-care is not an indulgence, but a necessity – a lifeline that staves off burnout and fuels resilience. As clinical psychologist Dr Julia Smith (2022) writes:

> In the balancing act of managing stress and using it to our advantage while remaining healthy, we need to balance incoming demands with replenishment. The more demands on us, the more replenishment we need. The more stress pouring into the bucket, the more release valves we need to process it and make room for the ongoing demands.
>
> (p. 248)

The narratives of everyday activists highlighted the power in establishing boundaries, finding solace in downtime, and embracing social support. Through adversity and triumph, everyday activists interviewed for this book gave us examples of how they embrace the cadence of rest, unapologetically establishing boundaries to safeguard their energy, and ensure their participation in causes they care about. To conclude this chapter, we suggest following the advice of feminist psychologist Nicola Jane Hobbs (2023): "Instead of asking, 'Have I worked hard enough to deserve to rest?' I've started asking, 'Have I rested enough to do my most loving and meaningful work?'"

References

Bakker, A. B., & Demerouti, E. (2007). The Job Demands-Resources model: State of the art. *Journal of Managerial Psychology*, 22(3), 309–328. doi: https://doi.org/10.1108/02683940710733115

Buckingham, M. (2008). *Go put your strengths to work: Six powerful steps to achieve outstanding performance.* Simon & Schuster UK.

Clifton, D. O., & Harter, J. K. (2003). Investing in strengths. In K. S. Cameron, J. E. Dutton, & R. E. Quinn (Eds.), *Positive organisational scholarship: Foundations of a new discipline* (pp. 111–121). Berrett-Koehler Publishers.

Deci, E. L., & Ryan, R. M. (1985). *Intrinsic motivation and self-determination in human behavior.* Springer Science & Business Media.

Deveson, A. (2004). Anne Deveson and resilience. Religion and Ethics ABC Broadcasting. https://www.abc.net.au/religion/watch/compass/anne-deveson-and-resilience/10142784

Galloway, S. (2019). *The algebra of happiness: Notes on the pursuit of success, love, and meaning.* Penguin.

Harper, F. (2020). *Unf*ck your boundaries: Build better relationships through consent, communication, and expressing your needs.* Microcosm Publishing.

Hobbs, N. J. (2023). *Nicola Jane Hobbs.* Facebook. https://www.facebook.com/NicolaJaneHobbs/

Kabat-Zinn, J. (1990). *Full catastrophe living: Using the wisdom of your body and mind to face stress, pain, and illness.* Delta.

Lazarus, R. S., & Folkman, S. (1984). *Stress, appraisal, and coping.* Springer.

Manson, M. (2017). *The subtle art of not giving a f*ck: A counterintuitive approach to living a good life.* Macmillan.

Maslow, A. H. (1943). A theory of human motivation. *Psychological Review, 50*(4), 370–396. doi: https://doi.org/10.1037/h0054346

Neff, K. (2003). Self-compassion: An alternative conceptualization of a healthy attitude toward oneself. *Self and Identity, 2*(2), 85–101. doi: https://doi.org/10.1080/15298860309032

Reid, E. M. (2012). Hanging up the cape: Boundary work, identity, and the end of a role. *Academy of Management Journal, 55*(2), 348–375.

Seligman, M. E. (2002). *Authentic happiness: Using the new positive psychology to realise your potential for lasting fulfillment.* Simon and Schuster.

Smith, J. (2022). *Why has nobody told me this before? Everyday tools for life's ups and downs.* Penguin Books.

Zahariades, D. (2021). *The art of saying NO: How to stand your ground, reclaim your time and energy, and refuse to be taken for granted (without feeling guilty!).* Independent.

9 Effective activism starts with everyday people

The initial impetus for writing this book was focused on the importance of evaluating activism. Our careers have been spent working in
non-profit organisations to strengthen their internal evaluation processes – we knew we had something that could be useful for activists
who wanted and needed to know if their efforts were making a difference. However, we soon discovered that some of our experiences as
activists in a more general sense may also be useful for readers and
linked back to our foundational degrees in philosophy and sociology.

Over time, the list of topics that we wanted to cover in the book
expanded and we started thinking about the people who were working
with us and who had inspired us over the years. These individuals were
our mentors, inspiration, and sources of wisdom. This led to the more
formal piece of empirical research that underpins this book. Utilising
our wide network of contacts across the non-profit and civil society
sector, we had 46 everyday activists agree to contribute via interview. It
was an inspirational part of the process to extract the gold nuggets of
knowledge and gems of wisdom from these tremendous individuals
who have maintained their efforts over the long term.

Although we had always been reflective about our efforts and read
widely to understand more about being effective activists, we have grown
as a result of talking with the interviewees and exploring the literature in
more depth. We have experienced changes in mindset, tinkered with our
way of working, expanded how we work, and improved the effectiveness
of our actions. We feel more confident in our decisions, have less anxiety
around what we should or should not be doing, and know more about
how to handle our reactions toward others.

Prioritising self-care has been one of our behaviours that has significantly changed. Following the interviewee examples of self-apprecia-
tion, we have spent more time showing ourselves gratitude for our
contribution. Buying flowers, experimenting with float therapy,

DOI: 10.4324/9781003333982-9

intentionally scheduling social catch ups, appreciating small moments of joy, being adventurous, conducting breathing exercises, absorbing compliments, and spending more time outdoors, are just some examples of changes we have made. Rather than feeling guilty about taking time out, writing Chapter Eight, "Self-care", has qualified the importance of this undeniably necessary factor for sustaining longevity and effectiveness. Reinforcing boundaries to safeguard time and space for contemplation has now become a routine way of *recharging the batteries*. When things have become too much, we recognised the feelings of guilt at not being able to do everything we set out to do, but ended up choosing a more rational path of reducing the workload to ensure a more appropriate balance. Similarly, as we have learnt new skills and attempted to apply the findings from Chapter Six on "Dealing with people who disagree", we have been buoyed by the recognition of our own courage. When facing challenging interpersonal situations, we acknowledge both how worthwhile it is to engage and how much energy is expended. In essence, writing this book has helped us understand the personal toll, value how much energy needs to be replenished, and acknowledge our actions as part of a personal growth journey.

The most important change from writing this book, however, is the liberating experience of overcoming feelings of paralysis. As evaluators by trade, we spend lots of time in our heads working on diagrams and theories of all the potential outcomes and repercussions of our actions. We also analyse the underlying assumptions and overlay the different viewpoints of stakeholders before dreaming up what we think the eventual result might be! Recognising the complexity of systems and the interrelatedness of our actions carries a high risk of overthinking things. Although we value these approaches for interrogating the situation, we found that writing Chapter Two, "Facing a problematic world", helped us consolidate the balance between acknowledging the structural forces at a societal level and the power of our micro-actions on an everyday level.

Hence, it was liberating to find evidence for and write about how our everyday micro-actions can and have contributed to major social change. We are part of the problem and part of the solution. As such, we recognise our part in contributing to issues while simultaneously embracing our autonomy and taking control. We can be people who understand our privilege and use our power to do something about the issues we care about. From that critical realisation stems the additional approaches and ways of working that can enhance our effectiveness. We are now taking more joy in harnessing our leverage to make a small but significant contribution to the issues that matter most to us.

We hope that readers also found strength and practical support in these pages. Just as we loved speaking with the inspirational interviewees, we hope their stories and experiences throughout the book similarly resonated and provided guidance. The purpose of this conclusion is to summarise the key takeaways so the reflective prompts are accessible. Part of being an effective activist is staying informed and adaptable; thus, returning to these topics and prompts as new research emerges will support personal growth and a constantly evolving approach. We suggest that everyday activists may find value in Layla Saad's (2020) recommendation to write down your responses to the questions. Articulating your answers, reflecting upon them after a period of time, and adding to the reflections as things change, can be a useful process. As such, this book can provide a framework for your ongoing reflections as experiences occur and new evidence emerges.

Key takeaways

Facing a problematic world

Understandably, we can feel despondent and paralysed by the enormity of the global situation. Becoming apathetic, disconnecting, or choosing to be ignorant offers no benefit to either ourselves or the world. Recognising that we are complicit in the causation and maintenance of oppressive systems is also a hard fact to accept. However, recognising how we fit within structural systems is a liberating step to determine what we can achieve by pooling our everyday actions on an individual level. Drawing from systems thinking and sociology, Chapter Two focused on the importance of taking action, despite our limitations and transient existence. It provided some strategies for how to deal with feelings of inertia. Some people make a choice to focus on what they can control. Alternatively, others focus on the future and the big-picture goal and see themselves in a historical context. Interviewees also suggested that reorienting thought patterns to focus on wins was a way of remaining positive. Others used anger and frustration as a motivational force or identified challenges as opportunities for learning by adopting a growth mindset. Reflecting on the emotional responses and understanding our personal triggers form the initial step that will help to work out what will move us from paralysis towards recognising our everyday power, helping to dismantle the structural systems, and joining up with networks for collective change.

The strategies for overcoming inertia and paralysis mentioned in Chapter Two that could be used in any combination and at different times include:

- Take time to identify and think about the inherent structures that prevent or hinder change.
- Identify your sphere of influence and focus on a manageable locus of control.
- Focus on the goal and decide what side of history you would like to be on, noting that often silence is complicity.
- Treasure instances of change and positive feedback no matter how small.
- Channel feelings of being overwhelmed, angry, and frustrated into momentum for action.
- Understand mistakes and barriers as learning opportunities for growth.
- Identify personal triggers for demotivation and avoid them or take a break.
- Draw upon internal belief and social systems for support.

Exploring underlying motivations

Exploring the motivations, values, and beliefs for why activists are propelled from inertia into action provides insight into where we should be placing our efforts. Understanding motivations can help us find fulfilment and identify what makes life meaningful. Being able to articulate a response to the question, "*Why do you do what you do?*" is not only useful when replying to others but can be a tool to affirm that you are on the right path or reassess and change course so that your actions align more closely with your motivations. People interviewed for this book could not simply ignore injustice or unfairness. Some were motivated by past or present experiences, the attainment of new knowledge, a need to live in accordance with values, and a desire for justice. Others were bound by duty or faith, feelings of guilt or empathy, or were influenced by their family and desire to leave a legacy. The philosophical and psychological theories drawn upon in Chapter Three may help readers determine whether their actions help give their lives meaning. We argue that meaningfulness exists when a person can find something they can do and enjoy doing, and these efforts are applied to a project that is worthwhile.

The reflective questions below are crafted to help identify motivations and connections between where you find your motivation, whether your actions have value beyond yourself, and what you find fulfilling:

Motivation

- What motivates you into activism?
- Are there multiple motivations that intersect?
- How have your motivations changed over time?
- Which actions are driven by an intrinsic passion?
- What would success look like for you and how important is it that you achieve success?

Application to something larger than oneself

- How are you creating, promoting, protecting, honouring, preserving, affirming, adding something?
- In what ways are the object of your attention worthy of your skills, passion, devotion, and time?
- Can you identify where your actions have an affect beyond your self-interest?
- How does your activism align with your moral code, or religious worldview, or something else?

Fulfilment

- How does activism make you feel?
- What do you enjoy doing?
- What activities are you passionately engaged with or find gripping?
- What sacrifices do you have to make to engage in activism?
- What experience, skills, and attributes can you offer?

Finding fulfilment

Dedicating time to activism can result in a wide range of emotional responses. Chapter Four interrogated the concept of fulfilment in more depth to promote the idea that developing self-awareness of personal effects and developing a stronger sense of personal identity can help people contribute more constructively for longer. Interviewees stated that they could sometimes feel like an insignificant imposter and easily become deflated and depleted. Other wide-ranging sensations included feeling a sense of purpose, utility, and even euphoria and gratification. Most interviewees were quick to point out that there were many social benefits that arise from working with others towards a similar goal. We argued that unpacking these emotional responses can help determine where we should be placing our efforts and where we need to develop and grow on a personal level. Activism can help us work towards self-actualisation and become better versions of ourselves.

These questions prompt reflection on self-awareness and identifying a path towards the liberating state of self-actualisation:

- How does my activism work contribute to my long-term goals and aspirations?
- To what extent do I feel a sense of fulfilment and purpose through my activism?
- What steps can I take to align my activism more closely with my self-actualisation goals?
- How does my activism work resonate with my authentic self?
- In what ways has my activism work evolved over time?
- What skills and knowledge have I gained through my activism?
- What challenges have I faced in my activism journey and how have they shaped my resilience and problem-solving skills?
- What have I learned from collaborating with others in activism?

Cooperating with others

Increasing the impact of your efforts will inevitably mean encouraging other people to get involved. Creating a supportive environment for cooperation to occur can mean a more effective and productive team. Chapter Five drew on social psychology to illustrate how to intentionally establish a cooperative team that can amplify achievements and ease the burden of a heavy workload. Acknowledging interdependence of networks and finding mutually beneficial situations were key themes. Forming meaningful connections with others is a skill that can be taught. Research can help us learn how to recruit people into a group using an individualised approach so that their contribution is valued. Activists can support positive group dynamics by establishing shared goals, providing encouragement, being inclusive, holding each other accountable, and reflecting on how the group is functioning (Johnson & Johnson, 2015). Conflict is inevitable in activism as people bring their different worldviews, value systems, and styles of working. However, activists can develop conflict management skills, just like communication, facilitation, and cultural competence skills, and learn how to avoid conflict, prevent it from escalating, and deal with it appropriately when it does arise. Activists do not have to hope for great team dynamics – they can instead think intentionally and strategically to decrease stress and enhance the likelihood of cooperative, effective, and uplifting group dynamics.

These questions, placed against each element of the social psychological recipe for positive team dynamics are prompts to support development of well-functioning teams:

Establishing shared goals

- Have you clearly articulated your goal?
- Do team members agree on what they are working towards?
- Does the team have long and short-term goals and descriptions of what success would look like?
- Have you collaboratively developed a plan that connects individual roles to the goals?

Providing encouragement

- How does the group provide recognition to individual members?
- Does the team trust each other enough to provide honest feedback?
- Are milestones celebrated along the way?
- Does the team have opportunities to engage in social settings?

Developing social skills

- How are different styles of communicating, learning, and engaging incorporated?
- How do you welcome and encourage diversity?
- Is there an agreed way of dealing with conflict when it arises?
- Is information communicated clearly and tailored for group members?

Being accountable

- What mechanisms do you use to hold yourself and others accountable?
- Does the group have ways of following through on commitments?
- Does the group have a system to manage information?
- Do agendas, minutes, or communication logs make decisions transparent and accessible?

Reflecting on the process

- Do you take time to consider whether the group is functioning well?
- Do you have a process for finding out from group members what they think?
- Are there scheduled opportunities for reflection?
- Do decision making processes include a feedback mechanism?

Dealing with people who disagree

Energy can be wasted dealing with people who may never agree with your cause. However, in acknowledging that activists will always encounter people who disagree or who are in denial, Chapter Six focused on helping readers determine when to walk away and when to engage. A key message was that listening is an important skill for allowing us to hear alternative viewpoints and identify our own biases. Recognising that it takes courage and confidence to participate in challenging discussions, we shared examples of how our interviewees removed themselves from inflammatory situations, used listening skills to form personal connections, and effectively challenged others on their views when appropriate. We suggested that it is worthwhile for activists to claim their stance in the long-term anti-denialism campaign and use inter-personal strategies to help change the attitudes and behaviours of people within our trusted social circles. However, more importantly, we argued that to develop a tolerant society we need to develop the skills required to listen to diverse opinions openly and authentically without getting defensive and build our capacity for addressing arguments without disengaging.

Psychologists Alison and Alison (2020) suggested the following reflection prompts to help us identify areas for growth to hone our ability to tailor interactions for mutually beneficial outcomes:

- Am I being honest or am I trying to manipulate the other person?
- Am I being empathic and seeing things from their perspective or just concentrating on my own point of view?
- Am I respecting and reinforcing their autonomy and right to choose, or am I trying to force them to do what I want?
- Am I listening carefully and reflecting to show a deeper understanding and build intimacy and connection? (Alison & Alison, 2020, p. 146)

Everyday evaluation

Chapter Seven emphasised the value of activists using evidence-based strategies to maximise their effectiveness. Research findings and other sources of information can help activists better understand what needs to be done, help articulate why they are doing what they are doing, and guide their actions. The chapter was about reflecting, tracking, and sharing acti-vist efforts. Everyday evaluation was highlighted for its value toward improving and learning through informal and continuous ways. We shared actionable guidance around setting goals, identifying outcomes, gathering data, and understanding the value of actions. We encouraged activists to understand how worthwhile it is to condense their insights and lessons into content that can be shared strategically. Translating this information for a broader audience can itself be a powerful advocacy tool.

Tracking, reflecting, and sharing activist work needs to be tailored to the context so that the method or approach provides relevant data and corresponds with available resources. The impact of activist efforts may be extended by incorporating some of these monitoring, evaluation, and learning strategies:

- Gather information to determine whether your approach is based in evidence.
- Develop a simple plan for tracking activist work.
- Consider whether you want to draw upon an existing approach or model to develop key questions.
- Collect and store relevant data to measure progress towards outcomes.
- Seek feedback through engaging with peers, mentors, or community members to gather external perspectives.
- Carve out space and time for reflective practice. Maintain a reflective journal or digital record to capture insights, challenges, and evolving perspectives.
- Simplify data by producing clear visualisations, infographics, or plain language summaries.
- Craft narratives and stories that highlight your real-world impact.
- Tailor messages to resonate with your audience's values, beliefs, and interests.
- Involve community members, recipients, partners, and other stakeholders in sharing your impact.
- Leverage diverse platforms such as social media, workshops, and community events to disseminate information.

Self-care

The preservation of one's own wellbeing is not just a necessity, it is also a catalyst for enduring and effective activism. Drawing on psychological theories, Chapter Eight encouraged readers to cultivate proactive strategies that foster resilience, prevent burnout, and promote personal growth so that a commitment to activism can be sustained long term. Most interviewees were aware of burnout as a serious mental health problem and had experienced feelings of being overwhelmed, overworked, and fatigued. We argued that acknowledging burnout as a legitimate threat is the first step towards adopting comprehensive and proactive self-care measures to mitigate its risks.

This list summarises the array of strategies offered by interviewees for promoting and safeguarding wellbeing because they recognised the significance of self-care:

Ensure needs are met

- Make a commitment to look after yourself and prioritise personal health.
- Engage in activities that ignite joy and foster a sense of rejuvenation.
- Cultivate strong connections with friends, family, and mentors for emotional sustenance.
- Create a self-care routine that prioritises and takes control of your wellbeing. Incorporate a diverse mix of activities that uplift your spirit, ensuring a well-rounded approach to wellbeing.

Reframe challenges

- Develop a habit of reframing challenges into opportunities for growth. Consciously shift your perspective to see the potential lessons in adversity.
- Enhance self-awareness by documenting stressors and how they have contributed to your personal development.
- Nurture a mindset of resilience and growth by integrating gratitude into your daily routine.

Utilise personal strengths

- Tailor your activism to align to your unique skills, qualities, and traits.
- Team up with activists who possess complementary strengths to create a blend of talents.
- Enhance your strengths through courses, mentorship, or self-guided learning.
- Identify and embrace your limitations to find roles that accommodate your comfort zone, allowing you to flourish.

Assess available resources

- Identify the resources at your disposal – time, skills, assets, financial support, and networks – to aid efficient allocation and prevent resource overuse.
- Stocktake the demands of your activism efforts to gauge if demands and resources are balanced.
- Break down goals into smaller manageable tasks.

- Recognise gaps and seek resources or support through encouraging sharing with other activists.
- Collaboratively develop a delegation protocol, distributing responsibilities among a team.

Maintain self-determination

- Define your personal values and the causes that resonate with you.
- Set achievable goals that align with your passions, giving you a sense of control and ownership.
- Identify the aspects of activism that bring you joy and fulfilment. Focus on activities that align with your intrinsic passions. Embrace what truly motivates you to ensure that your efforts remain authentic and sustainable.
- Seek out like-minded individuals or groups who share your passions and values to create a network that fosters a sense of belonging and solidarity.

Exercise mindfulness and self-compassion

- Infuse mindfulness into your daily routine. Engage in tasks with focused attention to create opportunities for present-moment awareness that help manage stress and foster a sense of calm.
- Practice deep breathing exercises and meditation to anchor yourself in the present moment.
- Engage in kind and compassionate self-talk to counter self-criticism and perfectionism.
- Open up to trusted friends, family, fellow activists, or access professional debriefing and counselling services for a structured platform to process challenges and emotions.

Set clear boundaries

- Define specific working hours and spaces for activism-related tasks. Once those hours are over, switch off from work-related communications and activities to focus on personal time.
- Openly communicate your boundaries with colleagues, family, and friends. Establish clear guidelines for communication channels and response times.
- Learn to say no. Practice assertiveness by declining additional commitments without guilt, emphasising the need to maintain a balanced workload.

- Raise awareness about the significance of boundary management within your activist community. Encourage others to engage in respectful boundary-setting to foster a culture of wellbeing.

Navigating a way forward

This section presents our suggestions for navigating the future. By no means have we mastered these issues, but we share in the spirit of recognising the importance of these critical factors. We include examples of how adopting a learning mindset, listening carefully, continually developing self-awareness, and slowing down can help everyday activists to be more effective over the long term. We advocate for the uptake of these suggestions so that ultimately, we can all upscale our efforts and enhance our contributions toward saving the planet.

Adopt a learning mindset

Activism operates in a dynamic space and often requires adaptation and evolution. Seek out new research and information, stay open to feedback, make time to reflect, continually educate yourself, and be willing to pivot when necessary. Celebrate successes, but also view failures or setbacks as opportunities to grow and refine your approach. By constantly learning and iterating, effectiveness increases over time and you will find deeper satisfaction in the journey, knowing you are always evolving toward a better version of your activist self.

Being an effective activist does not mean doing things the same things you have always done. New evidence will open up opportunities for doing things differently. These innovations or even slight adjustments may mean your actions reach more people, can be achieved in a shorter time period, use less resources, leave you with more energy, or just allow you to look at things from a different perspective. Seeing the activist journey as a continual opportunity for growth means you can give yourself permission to change your course, articulate to others how you have learnt from mistakes, and adapt with minimal emotional distress when new information becomes available.

Listen

Many of the challenges everyday activists face can be overcome by listening. Most of the answers we need for how to be more effective can be found through taking on board what someone is trying to share. Most obviously, listening to others can help us understand more deeply what they are trying to say beyond a literal level. This includes both close friends who are trying to tell us something of value or people with whom we disagree but have a message we need to hear. Feedback from stakeholders linked to activist efforts, either as supporters or recipients, can provide phenomenal information to enhance effectiveness if we take the time to listen with open hearts and minds.

Just as important is listening to our internal dialogue, or the conversations we have with ourselves. Listening to what our bodies are telling us and reflecting upon how we feel physically and emotionally when we undertake certain tasks, can provide insight. Practicing reflexivity and mindfulness across all our actions and responses can help us understand that many of the most valuable lessons and sources of wisdom can be found within ourselves.

Continually develop self-awareness

Stay grounded in your *why* and understand why you are doing what you are doing. Remember the root cause or passion that drove you to activism in the first place. This emotional and moral anchor will sustain you during challenging times, ensure your actions align with your core beliefs, and provide a touchstone for authentic communication with others. When actions and efforts are tied to a deeply held belief or value, they resonate more profoundly with audiences and maintain personal motivation.

However, this does not mean that the underlying motivation cannot change over time. Being flexible enough to recognise that your effort or skills may be better placed working towards a different goal can be part of a mature reflexive process. There is value in developing the self-awareness to be able to articulate when change is required and for what reasons. As noted above, writing answers to reflexive questions or journalling can be a tool for monitoring personal growth. Reading widely and sourcing alternative opinions to identify where there is alignment or differences may help you to think on a different level. Finding safe spaces to have conversations with others who are prepared to challenge your opinions and prompt deeper reflection may also be useful.

Slow down

Slowing down can help everyday activists find opportunities for working in more effective, productive, and constructive ways. Activists often experience a feeling of urgency and a heightened sense of importance. Their passion and desire to make change can be a constant driver.

They often place unrealistic and unachievable expectations upon themselves in relation to goals and timeframes. The stress of such unobtainable goals creates its own microclimate that inhibits productivity and effectiveness.

Many of the chapters of this book contain strategies and recommendations that relate to the importance of slowing down. All chapters contain prompts for reflection, and by their very nature, reflective processes take time. If you do not have time to consider why you are doing what you are doing, to create the right conditions for supporting a cooperative environment, to listen attentively in a challenging conversation, to find evidence that you are making a difference, or to read, sleep, breathe deeply, or take care of yourself – then you are probably doing too much. Recognise that one person can only do what they can do. Consider your limits and consciously acknowledge your responsibility to take control over task management, time management, and your own happiness, balance, and sense of fulfilment. Nobody else is obligated to give you acknowledgement, express gratitude, or manage your workload. Slowing down and identifying how you can help yourself achieve more, with the finite resources you have available, are essential.

Conclusion

We started this book with a quotation from anti-racism activist Layla Saad (2020). She suggests that we should aspire towards becoming responsible custodians of our shared future, "The choice is yours. The moment is now. Help change the world. Become a good ancestor" (p. 184). Our aspirations centre on the commitment to leave this world in an improved state for ourselves and the generations that follow, and for all those, human and non-human, who are impacted by our collective existence. We base our actions on the understanding that we are interdependent with each other and everything on the planet. We try our best to bring hope and optimism to our cooperative interactions with others.

Concerted effort of every individual is required to dismantle the entrenched systems of oppression and marginalisation that have inflicted enduring harm across generations, species, continents, and oceans. Every contribution holds significance as we strive to forge a reimagined world. Each person possesses the potential to instigate transformative change. The consequences of everyday actions, whether consciously chosen or otherwise, ripple through the lives of all those with whom we intersect and through the legacies we shape. Our actions are within our control. The impact and outcome of the consequences are things we can reflect upon, track, and share with others to determine whether we are acting in the most effective way possible.

We stand at a pivotal juncture where we can either permit injustice to continue, or proactively elect to disrupt and disassemble problems within our sphere of influence. We recognise the significant role of everyday activists in contributing to a kinder, fairer, more peaceful world. Noting that our personal liberation is bound up with the liberation of all others, we implore everyday activists to consciously retain the responsibility of being a good ancestor. We can work towards making the world a better place whilst simultaneously growing, learning, and developing along a path towards self-actualisation. By helping others, we help ourselves. By freeing others, we free ourselves.

References

Alison, E., & Alison, L. (2020). *Rapport: The four ways to read people.* Penguin.

Johnson, D. W., & Johnson, R. (2015). Cooperation and competition. *International Encyclopedia of the Social & Behavioral Sciences*, 2(4), 856–861. doi: https://doi.org/10.1016/B978-0-08-097086-8.24051-8

Saad, L. (2020). *Me and white supremacy: How to recognise your privilege, combat racism and change the world.* Quercus.

Index

Printed in the United States
by Baker & Taylor Publisher Services